Integrated Mathematics 2

Teacher's Resources for Transfer Students

The Teacher's Resources for Transfer Students contains the Toolbox from *Integrated Mathematics 2* in blackline master format, relevant sections from the *Study Guide for Integrated Mathematics 1*, a correlation of the Toolbox Skills to sections in the textbook and to the *Study Guide* sections contained in this publication, Skills Inventories on various groups of units, the Objectives of the course, and answers to all exercises.

McDougal Littell
A Houghton Mifflin Company
Evanston, Illinois
Boston ◆ Dallas ◆ Phoenix

Contents

ISBN: 0-395-71858-9

23456789 - HS - 99 98 97 96 95

Helping Transfer Students

In today's mobile society, it is common for mathematics teachers to have students who move from one school to another during the school year. *Teacher's Resources for Transfer Students* for *Integrated Mathematics 2* has been designed to make it easier for you to help your transfer students make a smooth transition.

For students transferring into your class, this publication provides:

(1) a correlation of the *Integrated Mathematics 2 Toolbox* skills (essential skills assumed to have been taught in previous courses) to the sections in *Integrated Mathematics 2* where they are first needed and to relevant sections from the *Study Guide for Integrated Mathematics 1*;

(2) the *Toolbox* from *Integrated Mathematics 2* in convenient blackline master format to facilitate review of these skills;

(3) sections from the *Study Guide for Integrated Mathematics 1* that provide for further study of certain *Toolbox* skills;

(4) Skill Inventories, to help you evaluate which mathematical skills in *Integrated Mathematics 2* students already have and which skills are weak or unfamiliar;

(5) page references that key the answers to the Skill Inventory questions to the lessons in the *Study Guide for Integrated Mathematics 2* where the skill being tested is taught and practiced. If a student misses a question, you can easily refer him or her to the appropriate *Study Guide for Integrated Mathematics 2* pages.

For students transferring out of your class, this publication provides:

(1) a list of course objectives organized by units in *Integrated Mathematics 2* and

(2) a list of course objectives organized by mathematical strands. These lists can help you describe to a teacher in another school the content that a student in your class has completed.

Graphics Calculator Technology

If transfer students are not familiar with graphics calculators, you may wish to refer them to the *Technology Handbook* that begins on page 603 of the textbook. This handbook provides an overview of the major features of graphics calculators that are used in the course, plus exercises to strengthen understanding. Also available is *Using TI-81 and TI-82 Calculators*, a supplementary publication that provides an introduction to these calculators, together with programs and activities for students.

Interactive Learning

Transfer students may not have had opportunities to experience cooperative learning and exploratory activities. As a first step, you may wish to explain to transfer students the purpose of cooperative learning groups, how they work, what students' roles in them are, and how group activities are evaluated. A similar explanation with regard to exploratory activities may also be appropriate. A possible second step would be to have transfer students participate in a cooperative group as a recorder or checker. In this manner, transfer students can participate in a group activity in a way that allows them to observe how a group functions and what roles other students perform.

Correlation of Toolbox Skills

The Toolbox in *Integrated Mathematics 2* provides students with explanations of, and practice with, essential skills studied in earlier courses. The chart below correlates each Toolbox Skill to the place in the *Integrated Mathematics 2* course where the skill is first needed or used. Also included is a correlation to relevant sections in the *Study Guide for Integrated Mathematics 1*. To facilitate review of prerequisite skills, the Toolbox from the textbook is included in blackline master format starting on page 5. The *Study Guide* sections begin on page 39 with page references given in the chart.

Toolbox Skill Number	***Integrated Mathematics 1 Study Guide* Sections**	Needed Before Section
1. Using a Fitted Line	Scatter Plots (p. 39)	1-3
2. Using a Histogram and a Frequency Table	Histograms and Stem-and-Leaf Plots (p. 43)	1-4
3. Making a Stem-and-Leaf Plot	Histograms and Stem-and-Leaf Plots (p. 43)	1-1
4. Finding Mean, Median, Mode, and Range	Mean, Median, and Mode (p. 47)	1-4
5. Making a Box-and-Whisker Plot	Box-and-Whisker Plots (p. 49)	1-1
6. Finding Experimental Probability	Investigating Probability (p. 52)	1-2
7. Finding Theoretical Probability	Investigating Probability (p. 52)	6-3
8. Finding Geometric Probability	Geometric Probability (p. 56)	1-2
9. Using Order of Operations		2-4
10. Simplifying Algebraic Expressions		1-6
11. Expanding Products		1-6
12. Simplifying Expressions with Radicals		2-6
13. Solving Linear Equations		2-2
14. Solving Inequalities		7-1
15. Rewriting Equations and Formulas		2-3
16. Solving Square and Cube Root Equations		2-4
17. Solving Proportions		2-4
18. Using the Zero-Product Property		4-4
19. Recognizing Functions	Graphs and Functions (p. 59)	2-1
20. Making a Table of Values to Graph		2-2

Toolbox Skill Number	*Integrated Mathematics 1 Study Guide* Sections	Needed Before Section
21. Finding Slope		2-2
22. Modeling Direct Variation		2-2
23. Using $y = mx + b$		2-2
24. Using Intercepts to Graph		2-2
25. Graphing an Inequality in Two Variables		3-1
26. Finding Unknown Angle Measures	Exploring Angle Relationships (p. 63)	1-6
27. Reading Diagrams	Working Together on Congruent Polygons (p. 68); Exploring Quadrilaterals and Symmetry (p. 71)	1-5
28. Using Formulas from Geometry	Area Formulas (p. 75); Circumference and Arc Length (p. 81); Areas of Circles and Sectors (p. 84); Surface Area of Space Figures (p. 87); Volumes of Prisms and Cylinders (p. 92); Volumes of Pyramids and Cones (p. 96)	2-4
29. Finding Sine, Cosine, and Tangent Ratios	Sine and Cosine Ratios (p. 99); Direct Variation, Slope, and Tangent (p. 102)	8-8
30. Finding Missing Lengths in Right Triangles		8-7
31. Finding Missing Lengths		8-7
32. Determining a Right Triangle		8-7
33. Translating a Figure	Translations (p. 105)	3-6
34. Rotating a Figure	Rotations (p. 108)	5-4
35. Dilating a Figure	Dilations (p. 112)	3-6
36. Reflecting a Figure	Reflections (p. 115)	5-4

Name ______________________________ Date ______________

Toolbox

Skill 1 Using a Fitted Line

A **fitted line** is a line that passes as close to as many data points on a scatter plot as possible.

Example The table shows sales of CDs at a record store that was trying out different weekly schedules with different numbers of hours of operation. Make a scatter plot of the data. Draw a fitted line. Predict sales for a 45-hour week.

Hours of operation	54	48	60	65	40	60	48	56
CDs Sold	710	530	850	940	520	740	630	750

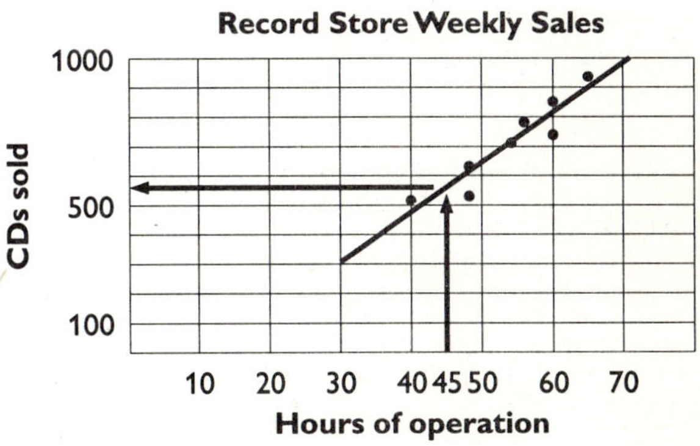

To predict sales for a 45-hour week, draw a vertical line from 45 on the horizontal axis to the fitted line. Then draw a horizontal line to the vertical axis. The point where the horizontal line meets the vertical axis is the prediction.

A 45-hour week should produce about 550 CD sales.

When two data sets increase together, their relationship is called a **positive correlation.** The points of a graph of the data *rise* to the right, as in the Example above. If the points of the graph of two data sets *fall* to the right, the graph shows a **negative correlation.**

PRACTICE

1. Make a scatter plot using the data in the table below showing the use of a small beach on days with different high temperatures.

High temperature (°F)	65	72	90	85	85	72	100	95
Beach Users	20	60	120	90	110	85	150	130

Name ______________________ Date __________

Skill 1 *(continued)*

2. Does your scatter plot show a *positive correlation*, a *negative correlation*, or *no correlation*?
3. Draw a fitted line for your scatter plot.

Use your fitted line to predict the number of beach users on days with each high temperature.

4. 80°F
5. 75°F
6. 105°F
7. 60°F

Skill 2 Using a Histogram and a Frequency Table

A **histogram** displays grouped data. Each group is defined by an interval on the horizontal axis. The height of the bar for an interval indicates the number of data items that fall in the group. This number is called a **frequency.**

Example The frequency table and the histogram show the number of supporters for candidate Kaoru Hirakawa in different age groups in her city.

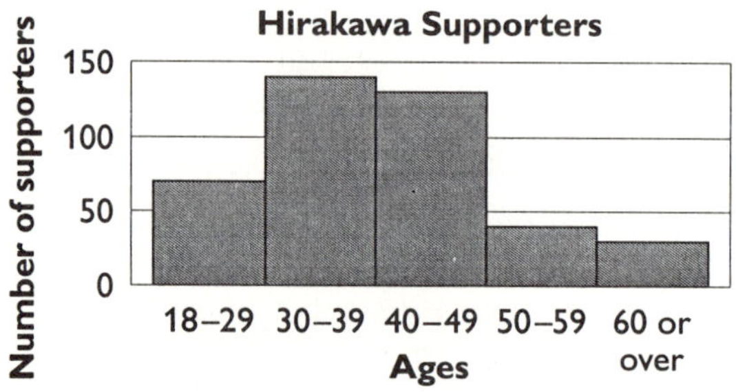

Age Group	Frequency
18–29	70
30–39	140
40–49	130
50–59	40
60 or over	30

Use the table or the histogram to answer each question.

a. How many Hirakawa supporters are between the ages of 50 and 59?

Histogram: The bar for the interval 50–59 has height 40.
Frequency table: The frequency for the age group 50–59 is 40.

There are 40 supporters between the ages of 50 and 59.

b. In which age groups does she have the greatest support and the least support?

Using the histogram, you see that the tallest bar is for the interval 30–39. She has the greatest support in this age group. The shortest bar is for the group "60 or over." She has the least support in this age group.

Name ______________________ Date ____________

Skill 2 *(continued)*

PRACTICE

Use the histogram or the frequency table to answer each question.

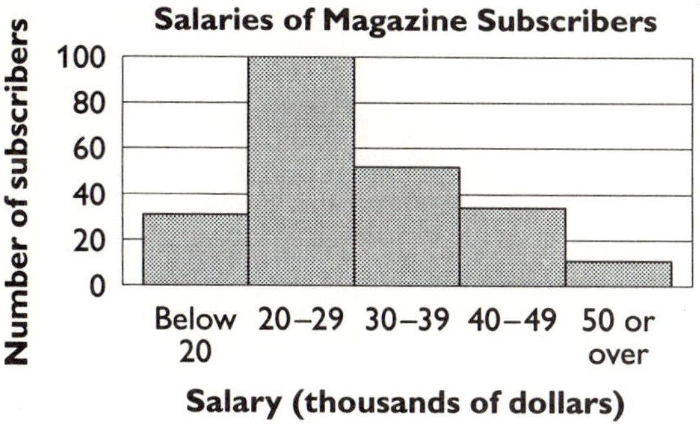

Salary Range (thousand $)	Number of subscribers
Below 20	30
20–29	100
30–39	52
40–49	34
50 or over	11

How many subscribers make each salary?

1. Below $20,000 **2.** Between $20,000 and $29,000 **3.** $50,000 or over

4. Which salary range has the most subscribers? the fewest?

Skill 3 Making a Stem-and-Leaf Plot

A **stem-and-leaf plot** displays every data item but also groups the data by intervals, like a histogram.

The leaf is the last digit on the right. → 1 5 2 ← The stem is the remaining digit or digits.

Example Make a stem-and-leaf plot from the table below that shows the daily protein intake of health club members for one day.

Protein Consumed (grams)				
63	68	69	82	84
43	84	87	70	72
52	75	71	68	44
56	57	63	76	64

Protein Consumed (grams)

Stem	Leaves
4	3 4
5	2 6 7
6	3 3 4 8 8 9
7	0 1 2 5 6
8	2 4 4 7

← This row displays the numbers 63, 63, 64, 68, 68, 69.

PRACTICE

1. In the Example, what data items do the numbers in the fourth row of the stem-and-leaf plot represent?

2. Make a stem-and-leaf plot of the data in the table at the right, showing times of runners in a 10 km road race.

3. In your stem-and-leaf plot, which interval contains the most data items? Which interval contains the fewest?

Finishing Times (minutes)				
52	34	45	63	71
45	41	32	50	62
52	44	43	51	46
51	66	65	66	58

Name ______________________________ Date ____________

Skill 4 Finding Mean, Median, Mode, and Range

mean: The sum of the data in a data set divided by the number of items.	**median:** The middle number or the mean of the two middle numbers in a data set when the data are arranged in order.
mode: The data item, or items, appearing most often. There may be more than one mode or no mode.	**range:** The difference between the smallest and largest data items.

Examples The following high temperatures were recorded on days in March:

42, 38, 56, 56, 60, 40, 65, 45, 48, 42, 56.

Find the mean, the median, the mode, and the range of the data.

Mean: There are 11 temperatures. Add the numbers. Divide the sum by 11.

$$\frac{42 + 38 + 56 + 56 + 60 + 40 + 65 + 45 + 48 + 42 + 56}{11} = 49.8$$

To the nearest tenth, the mean is 49.8.

Median: List the numbers in order. Find the middle number.

38 40 42 42 45 48 56 56 56 60 65

The median is 48.

Mode: Since 56 appears three times, and no other number appears more than twice, 56 is the mode.

Range: The smallest number is 38. The largest is 65. Subtract to find the range.

$$65 - 38 = 27$$

The range is 27.

PRACTICE

Use the table at the right showing protein per serving of some foods.

1. Find the mean of the data.
2. Find the median of the data.
3. Find the mode of the data.
4. Find the range of the data.

Protein per serving (g)					
24	36	42	10	32	24
10	18	16	18	12	30

Name ______________________ Date ____________

Skill 5 Making a Box-and-Whisker Plot

A **box-and-whisker plot** shows the median and range of data.

Example Make a box-and-whisker plot of these scores:
68, 78, 93, 79, 72, 88, 44, 79, 75, 82, 71, 89, 75, 88, 73, 80, 66, 100, 75, 92.

Step 1 Write the scores in order from lowest to highest.

Step 2 Find the median, the *extremes,* and the *quartiles* of the ordered data set.

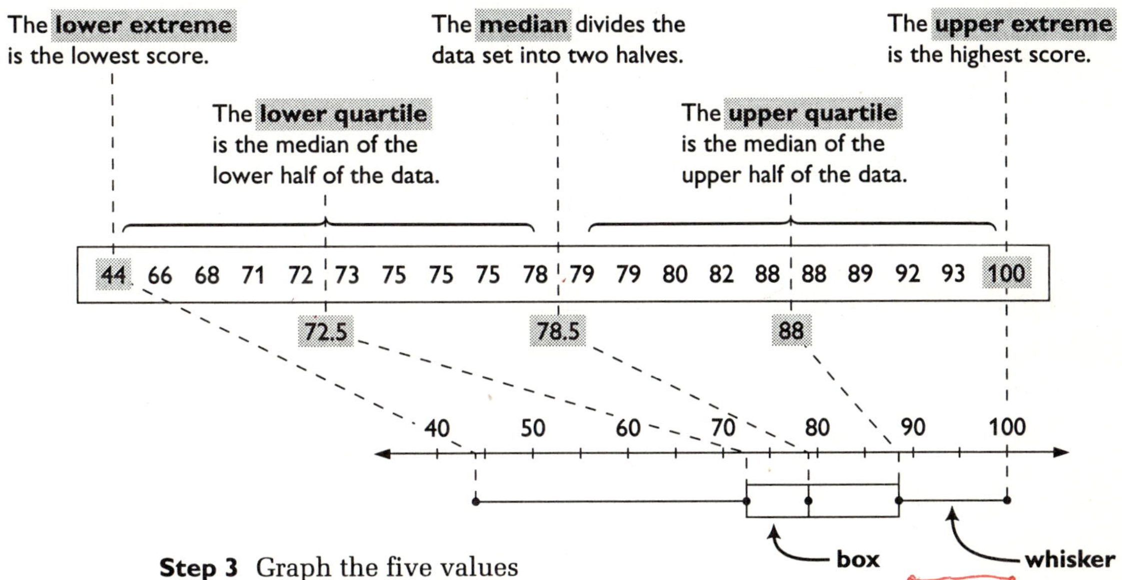

Step 3 Graph the five values below a number line.

Step 4 Draw a box from the lower quartile to the upper quartile. Draw a vertical line through the median.

Step 5 Draw line segments, or whiskers, from the box to the extremes.

You can see from the box-and-whisker plot that:

- About half the data are above 78.5.
- About 25% of the data are above 88.
- No data are below 44 or above 100.
- About 25% of the data are below 72.5.

PRACTICE

1. Make a box-and-whisker plot of the data shown in the table at the right.
2. For what number is it true that about 25% of the data fall above the number?
3. For what number is it true that about 25% of the data fall below the number?

Fuel Economy (mi/gal)				
34	28	36	38	28
30	40	38	26	22
32	34	32	44	42
48	20	22	26	30

Name ____________________ Date ____________

Skill 6 Finding Experimental Probability

To describe how likely an event is, you use a number on a scale from 0 to 1. This number is called the **probability** of the event.

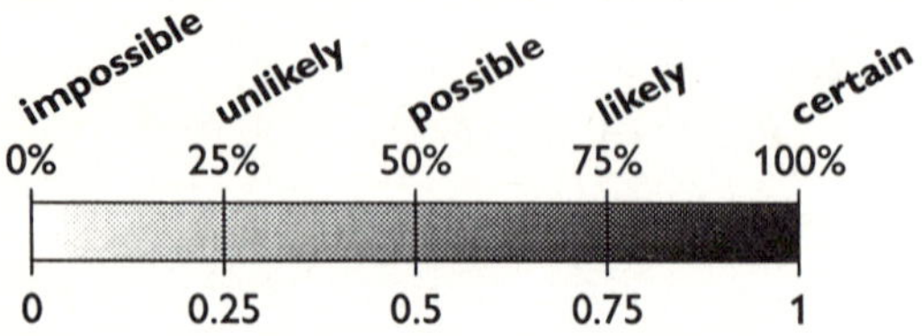

Probability that is based on the result of an experiment is called **experimental probability**. An **event** consists of one **outcome** or a group of outcomes of the experiment.

The experimental probability of an event E is the ratio:

$$P(E) = \frac{\textbf{number of times the event } E \textbf{ happened}}{\textbf{number of times the experiment was done}}$$

Example Janelle Franklin had 21 hits in her first 62 official at-bats for her high school softball team. What is the probability that she will get a hit?

The event E is "getting a hit."

$$P(E) = \frac{\text{hits}}{\text{at-bats}} = \frac{21}{62} \approx 0.339$$

The experimental probability of her getting a hit is about 0.34, or 34%.

Example Suppose Janelle has 20 more at-bats. How many hits can she expect?

Multiply the experimental probability by the number of future at-bats.

$(0.339)(20) = 6.78$

She can expect about 7 more hits.

PRACTICE

The table shows how many of each different size of a particular style of sweatshirt were sold at one store last month. Find the probability that a sweatshirt sold will be each size.

Size	Number Sold
Extra Small	12
Small	15
Medium	32
Large	44
Extra Large	14

1. Large

2. Medium

3. Extra Small

4. Extra Large

5. Suppose 150 sweatshirts are sold next year. How many will be Extra Large?

Name ______________________________ Date ______________

Skill 7 Finding Theoretical Probability

When all outcomes of an experiment are equally likely, you can find the **theoretical probability** of an event without performing any trials of the experiment. The theoretical probability $P(E)$ is found by counting the ways the experiment can turn out.

$$P(E) = \frac{\text{number of favorable outcomes}}{\text{number of possible outcomes}}$$ ⟵ "Favorable" means included in the event E.

Example A spinner like the one shown at the right is spun. What is the probability that the spinner ends up pointing at a number greater than 3?

The event E is that a number greater than 3 will come up.
There are five possible outcomes: 1, 2, 3, 4, or 5 comes up.
There are two favorable outcomes: 4 or 5 comes up.

$P(\text{number greater than 3}) = \frac{2}{5} = 0.4$

The theoretical probability is $\frac{2}{5}$, or 0.4, or 40%.

A standard deck of playing cards has 52 cards, with 13 cards in each of four suits: clubs, spades, diamonds, and hearts. Face cards are jacks, queens, and kings.

Example Suppose you pick a card at random from the 13 hearts of a standard deck. What is the probability that it is a face card?

There are 13 possible outcomes:

There are 3 favorable outcomes:

$P(\text{face card}) = \frac{3}{13} \approx 0.23$, or about 23%

PRACTICE

Use the spinner above. Find each probability.

1. The pointer lands on 5.

2. The pointer lands on an odd number.

3. The pointer does not land on 1.

4. The pointer lands on a number less than 6.

A card is chosen at random from the 13 hearts of a standard deck. Find each probability.

5. P(ace of hearts)

6. P(queen or king of hearts)

7. P(2, 3, or 4 of hearts)

8. P(not a face card)

9. P(a diamond)

10. P(not a diamond)

Name ______________________________ Date ______________

Skill 8 Finding Geometric Probability

Geometric probability is probability based on areas and lengths.

Example Suppose in a one-hour TV program, 18 minutes are taken up by commercials. You tune in at a random time. What is the probability that you will tune in on a commercial?

Step 1 Model the situation with a number line. Use minutes as units.

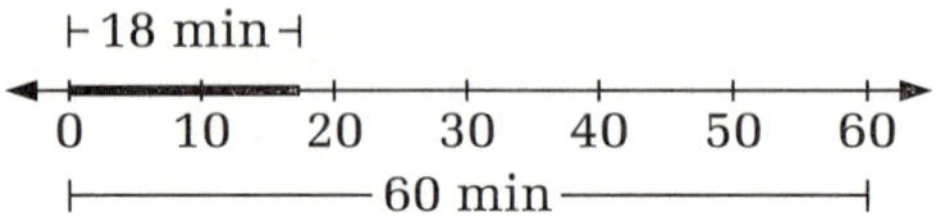

Step 2 Find the ratio of the length of the segment modeling commercial time to the length of the segment modeling total time.

$$P(\text{tuning in on a commercial}) = \frac{\text{commercial time}}{\text{total time}}$$

$$= \frac{18}{60} = \frac{3}{10}, \text{ or } 0.3, \text{ or } 30\%$$

If you can model the outcomes of an experiment by points in a plane, you can use geometric probability.

Example In a game, Trúc Ha hits a baseball over the park fence toward the wall of the house shown. Each window measures 2.5 ft by 4.5 ft. What is the probability that her hit will break a window?

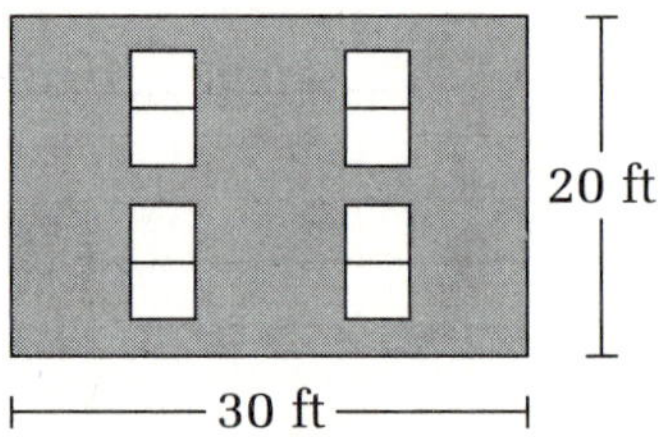

The area of the four windows is $4(2.5)(4.5) = 45$ (ft^2).

The wall area is $(20)(30) = 600$ (ft^2).

$$P(\text{breaking a window}) = \frac{\text{area of 4 windows}}{\text{wall area}}$$

$$= \frac{45}{600} = \frac{3}{40}, \text{ or } 0.075, \text{ or } 7.5\%$$

PRACTICE

1. In any 10-minute interval, a bus comes to the Eucalyptus Street stop and waits for 1.5 min. Suppose you arrive at the stop at a random time. What is the probability that a bus is waiting?

Find the probability that a point chosen at random in each figure is in the shaded region.

2. 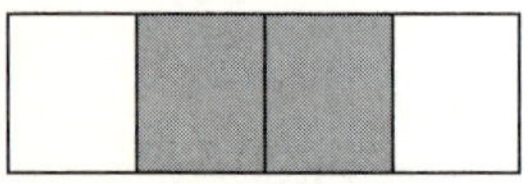

3. 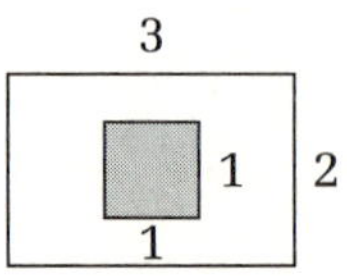

4. 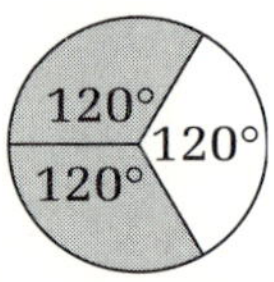

Name ______________________________ Date ______________

Skill 9 Using Order of Operations

People who work with mathematics follow a special **order of operations** so that they will agree on the value of a numerical expression. To simplify expressions such as $48 + (12 - 8)^2 \div 8$, you must carry out operations in the following order.

Parentheses	← Simplify inside parentheses.
Exponents	← Then calculate any powers.
Multiplication and Division	← Then do these as they appear from left to right.
Addition and Subtraction	← Then do these as they appear from left to right.

Example Simplify $48 + (12 - 8)^2 \div 8$.

$48 + (12 - 8)^2 \div 8 = 48 + 4^2 \div 8$ ← Simplify inside parentheses first.

$= 48 + 16 \div 8$ ← 4^2 means $4 \cdot 4$.

$= 48 + 2$ ← Do division next.

$= 50$ ← Do addition last.

Example Evaluate $7x^2 - 3x + 5$ when $x = -2$.

$7x^2 - 3x + 5 = 7(-2)^2 - 3(-2) + 5$ ← Substitute -2 for x.

$= 7 \cdot 4 - 3(-2) + 5$ ← $(-2)^2$ means $(-2)(-2)$.

$= 28 - (-6) + 5$ ← Do multiplication next.

$= 28 + 6 + 5$ ← To subtract -6, add its opposite.

$= 39$ ← Do addition last.

PRACTICE

Simplify each expression.

1. $12 \div 2^2 + 9$ **2.** $(13 - 5)^2 + 6 \div 2$ **3.** $4 + 5(7 - 9)$

4. $6 + 2^5 - 3 \cdot 5$ **5.** $7^2 - 4 \cdot 8 + 26 \div 13$ **6.** $(10 - 7)^4 + (5 - 1)^2 \div 8$

Evaluate each expression when $x = 3$.

7. $x^2 - 7x + 12$ **8.** $4x^3 - 5x + 1$ **9.** $15 - 4x^2$

10. $-2x^2 + 6x - 4$ **11.** $2x^5 - 4(x - 1)$ **12.** $(x - 5)^3 - 4(x + 1)^2$

Evaluate each expression when $a = -10$ and $b = 2$.

13. $3(a + b)$ **14.** $-a - b^2$ **15.** $-a^2b$

16. $-4a^2 + b^2$ **17.** $-4(a^2 + b^2)$ **18.** $-4(a + b)^2$

Name ______________________ Date __________

Skill 10 Simplifying Algebraic Expressions

You can simplify expressions like $3(x + 2)$ using the distributive property.

Distributive Property For all numbers a, b, and c,
$\boldsymbol{a(b + c) = ab + ac}$ and $\boldsymbol{a(b - c) = ab - ac}$.

Example $3(x + 2) = 3 \cdot x + 3 \cdot 2$
$= 3x + 6$

Example $5(a - 7) = 5 \cdot a - 5 \cdot 7$
$= 5a - 35$

To simplify some expressions, you use the fact that $-a = -1a$. Read $-a$ as "the opposite of a."

Example $-(y - 5) = -1(y - 5)$
$= (-1)y - (-1)(5)$ ← Use the distributive property.
$= -y - (-5)$
$= -y + 5$

Example $9 - 2(x + 4) = 9 - (2x + 2 \cdot 4)$ ← Use the distributive property.
$= 9 - (2x + 8)$
$= 9 - 2x - 8$ ← $-1(2x + 8)$
$= -2x + 1$

When you add expressions, each expression is a **term** of the sum. Terms with the same variables and powers are called **like terms.** You use the distributive property in reverse to **combine like terms.**

Example $2x^2 + 4xy - 5x^2 + 7xy = (2x^2 - 5x^2) + (4xy + 7xy)$ ← Group like terms.
$= (2 - 5)x^2 + (4 + 7)xy$ ← Use the distributive property.
$= -3x^2 + 11xy$

PRACTICE

Simplify.

1. $3(x + 4) - 5$

2. $-2(y - 7) + 3$

3. $4n - (3n + 5m)$

4. $5(-k + 3) - 2(k - 4)$

5. $4x^2 + 3x - x^2$

6. $4xy + 3yz - 7xy$

7. $-a^2 + ab - 5a^2$

8. $-2hk + 4k^2 - k^2 + 5hk$

9. $3 + 9x^3 - 10 - 2x^3$

Name ______________________________ Date ______________

Skill 11 Expanding Products

You can use the distributive property to expand a product.

Example $3x(x+2) = 3x \cdot x + 3x \cdot 2$
$= 3x^2 + 6x$

Example $-2x(x-5) = (-2x)x - (-2x)5$
$= -2x^2 - (-10x)$
$= -2x^2 + 10x$

Example $(x+4)(x-3) = (x+4)x - (x+4)3$ ← Use the distributive property.
$= x^2 + 4x - (3x + 12)$ ← Use the distributive property again.
$= x^2 + 4x - 3x - 12$
$= x^2 + x - 12$

PRACTICE

Expand each product.

1. $x(2x+3)$
2. $-4x(x-5)$
3. $7a(-2a+6)$
4. $(x+5)(x-9)$
5. $(y-6)(y-2)$
6. $(x-3)(-x+10)$
7. $(2x+1)(x-7)$
8. $(-3n+7)(n+5)$
9. $(4x-2)(x-5)$
10. $(3x-4)(2x-11)$
11. $(2x+1)(7x-8)$
12. $(-3k+2)(-5k+4)$

Skill 12 Simplifying Expressions with Radicals

To simplify the square root of an integer, you first find the largest perfect-square factor of the integer. Then use the fact that for all nonnegative numbers a and b,

$\sqrt{\ }$ is the radical symbol. → $\sqrt{ab} = \sqrt{a} \cdot \sqrt{b}$.

Example $\sqrt{18} = \sqrt{9 \cdot 2}$ ← 9 is the largest perfect-square factor of 18. Write 18 as $9 \cdot 2$.
$= \sqrt{9} \cdot \sqrt{2}$ ← Use the fact that $\sqrt{ab} = \sqrt{a} \cdot \sqrt{b}$.
$= 3\sqrt{2}$ ← Write $\sqrt{9}$ as 3.

Note: It would *not* have been helpful to write $\sqrt{18}$ as $\sqrt{6 \cdot 3}$, because neither 6 nor 3 is a perfect square.

Name ______________________ Date ____________

Skill 12 *(continued)*

To simplify a product of radical factors, you use the fact that for all nonnegative numbers *a* and *b*,

$$\sqrt{a} \cdot \sqrt{b} = \sqrt{ab}.$$

Example $5\sqrt{3} \cdot \sqrt{2} = 5\sqrt{3 \cdot 2}$ ← Use the fact that $\sqrt{a} \cdot \sqrt{b} = \sqrt{ab}$.

$= 5\sqrt{6}$

Example $4\sqrt{5} \cdot 2\sqrt{15} = 4 \cdot 2 \cdot \sqrt{5} \cdot \sqrt{15}$ ← Group radical terms.

$= 8\sqrt{5 \cdot 15}$ ← Use the fact that $\sqrt{a} \cdot \sqrt{b} = \sqrt{ab}$.

$= 8\sqrt{75}$ ← Simplify.

$= 8\sqrt{25 \cdot 3}$ ← 25 is the largest perfect square factor of 75.

$= 8\sqrt{25} \cdot \sqrt{3}$ ← Use the fact that $\sqrt{ab} = \sqrt{a} \cdot \sqrt{b}$.

$= 8 \cdot 5\sqrt{3}$ ← Write $\sqrt{25}$ as 5.

$= 40\sqrt{3}$

PRACTICE

Simplify.

1. $\sqrt{45}$ **2.** $\sqrt{300}$ **3.** $\sqrt{76}$ **4.** $5\sqrt{12}$ **5.** $12\sqrt{125}$

6. $\sqrt{6} \cdot \sqrt{7}$ **7.** $3\sqrt{5} \cdot \sqrt{2}$ **8.** $4.2\sqrt{11} \cdot \sqrt{3}$ **9.** $8\sqrt{2} \cdot 3\sqrt{3}$

10. $\sqrt{6} \cdot \sqrt{6}$ **11.** $(3\sqrt{3})^2$ **12.** $2\sqrt{6} \cdot 3\sqrt{2}$ **13.** $4\sqrt{5} \cdot 3\sqrt{10}$

Skill 13 Solving Linear Equations

You can solve some equations by *undoing* operations. Your goal is to get the variable alone on one side of the equation.

Example Solve $\frac{x}{2} + 7 = 40$.

$\frac{x}{2} + 7 - 7 = 40 - 7$ ← Undo addition of 7. Subtract 7 from both sides.

$\frac{x}{2} = 33$

$2 \cdot \frac{x}{2} = 2 \cdot 33$ ← Undo division by 2. Multiply both sides by 2.

$x = 66$

Name ________________________________ Date ____________

Skill 13 *(continued)*

Example Solve $\frac{n-5}{3} = -11$.

$$3 \cdot \frac{n-5}{3} = 3(-11) \quad \leftarrow \text{Multiply both sides by 3.}$$
$$n - 5 = -33$$
$$n - 5 + 5 = -33 + 5 \quad \leftarrow \text{Add 5 to both sides.}$$
$$n = -28$$

When the variable appears on both sides of an equation, you will need to get the variable terms on the same side.

Example Solve $7x = 3(x + 1)$.

$$7x = 3x + 3 \quad \leftarrow \text{Use the distributive property.}$$
$$7x - 3x = 3x + 3 - 3x \quad \leftarrow \text{Subtract } 3x \text{ to get the variable terms on one side.}$$
$$4x = 3 \quad \leftarrow \text{Combine like terms.}$$
$$x = \frac{3}{4}, \text{ or } 0.75 \quad \leftarrow \text{Divide both sides by 4.}$$

Example Solve $\frac{2}{3}x = 4x - 10$.

$$\frac{2}{3}x - 4x = 4x - 10 - 4x \quad \leftarrow \text{Subtract } 4x \text{ from both sides.}$$
$$\frac{2}{3}x - 4x = -10$$
$$\frac{3}{2}\left(\frac{2}{3}x - 4x\right) = \frac{3}{2}(-10) \quad \leftarrow \text{Multiply both sides by } \frac{3}{2}\text{, the reciprocal of } \frac{2}{3}.$$
$$x - 6x = -15 \quad \leftarrow \text{Use the distributive property.}$$
$$-5x = -15 \quad \leftarrow \text{Combine like terms.}$$
$$x = 3 \quad \leftarrow \text{Divide both sides by } -5.$$

PRACTICE

Solve each equation.

1. $\frac{x}{6} = 12$
2. $\frac{n}{2} + 5 = 8$
3. $\frac{3}{4}x - 11 = 4$
4. $\frac{x+5}{3} = 10$
5. $\frac{x-7}{2} = 2$
6. $\frac{n-15}{2} = -5$
7. $\frac{w}{10} - 3.2 = 5.4$
8. $\frac{x-7}{2} + 4 = 2$
9. $3x = 2x - 5$
10. $-2x + 35 = 3x$
11. $\frac{x-7}{2} + 4 = 2x$
12. $5(x - 8) = 2x + 8$
13. $\frac{x}{3} - 5 = -4x$
14. $\frac{2}{3}n - 7 = n$
15. $\frac{2x+1}{5} = x$

Name ______________________________ Date ______________

Skill 14 Solving Inequalities

To solve an inequality, you can use the same "undoing" operations that you used to solve equations.

Example Solve and graph the inequality $2x - 3 < 5$. Read "is less than."

$$2x - 3 < 5$$

$$2x - 3 + 3 < 5 + 3 \quad \leftarrow \text{Undo subtraction of 3. Add 3 to both sides.}$$

$$2x < 8$$

$$\frac{2x}{2} < \frac{8}{2} \quad \leftarrow \text{Undo multiplication by 2. Divide both sides by 2.}$$

$$x < 4$$

Draw a ray to the left to include all points less than 4.

Use an open circle to show that 4 is not included.

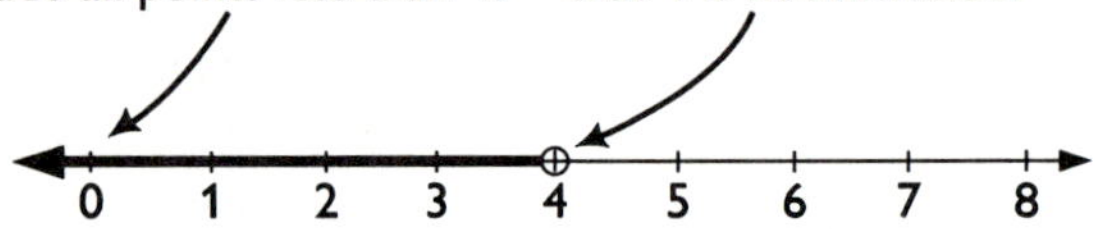

An inequality is reversed when you multiply or divide by the same negative number on both sides of an inequality.

Example Solve and graph the inequality $-2(x + 1) \leq 8$. Read "is less than or equal to."

$$-2(x + 1) \leq 8$$

$$\frac{-2(x + 1)}{-2} \geq \frac{8}{-2} \quad \leftarrow$$ When you divide both sides by -2, reverse the inequality symbol. Read "$\geq$" as "is greater than or equal to."

$$x + 1 \geq -4$$

$$x + 1 - 1 \geq -4 - 1 \quad \leftarrow \text{Subtract 1 from both sides.}$$

$$x \geq -5$$

Use a closed circle to show that -5 is included.

Draw a ray to the right to include all points greater than -5.

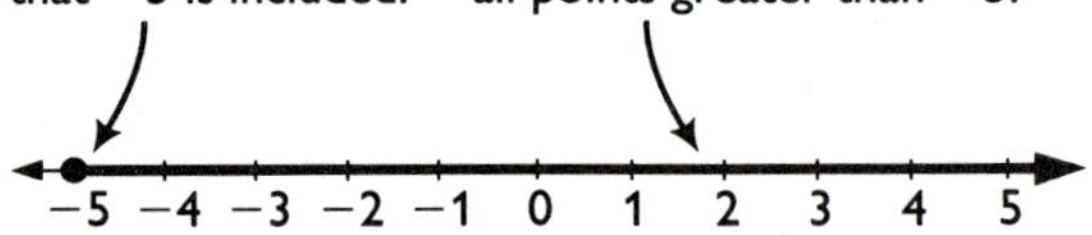

PRACTICE

Solve and graph each inequality.

1. $x + 3 < 7$ **2.** $2 - x \leq 4$ **3.** $-3x < 15$

4. $2(x - 1) \geq -6$ **5.** $10 > -5(x + 2)$ **6.** $-2x + 7 \leq 5$

7. $4 - 3x > -2$ **8.** $5(4 - x) \geq 20$ **9.** $x > 3x + 8$

Name ______________________________ Date ______________

Skill 15 Rewriting Equations and Formulas

Solving a formula or an equation for a variable means rewriting the formula or equation to get the variable alone on one side.

Example Solve the formula $A = lw$ for w.

$A = lw$

$\frac{A}{l} = \frac{lw}{l}$ ← Undo multiplication by l. Divide both sides by l.

$\frac{A}{l} = w$

Example Solve the equation $3x + 2y = 10$ for y.

$3x + 2y = 10$

$3x + 2y - 3x = 10 - 3x$ ← Undo addition of $3x$. Subtract $3x$ from both sides.

$2y = 10 - 3x$

$\frac{1}{2} \cdot 2y = \frac{1}{2}(10 - 3x)$ ← Multiply both sides by the reciprocal of 2.

$y = 5 - \frac{3}{2}x$ ← Use the distributive property.

PRACTICE

Solve each equation for the variable shown in bold.

1. $A = \frac{1}{2}b\mathbf{h}$
2. $C = 2\pi\mathbf{r}$
3. $A = \frac{1}{2}(a + \mathbf{b})$
4. $2x + \mathbf{y} = 13$
5. $5x + 2\mathbf{y} = 8$
6. $m = (\mathbf{n} - 2)180$
7. $2\mathbf{x} + y = 13$
8. $y = m\mathbf{x} + b$

Skill 16 Solving Square and Cube Root Equations

You can "undo" the squaring of a variable in an equation by taking the square root of both sides. Remember that every positive number has both a positive and a negative square root. The symbol $\sqrt{3}$ means "the *positive* square root of 3."

Example Solve $x^2 = 14$.

$x^2 = 14$

$x = \pm\sqrt{14}$ ← Take the square root of both sides. Use "±" to show that 14 has both a positive and a negative square root.

$x \approx \pm 3.74$

Name ______________________________ Date ______________

Skill 16 *(continued)*

Example Solve $x^2 + 16 = 81$.

$$x^2 + 16 - 16 = 81 - 16$$ ← Undo addition of 16. Subtract 16 from both sides.

$$x^2 = 65$$

$$x = \pm\sqrt{65}$$ ← Take the square root of both sides.

$$x \approx \pm 8.06$$ ← 65 INV x^2

You can undo the cubing of a variable in an equation by taking the cube root of both sides. Remember that a positive number has a positive cube root, and a negative number has a negative cube root. There is no "±" in the solution.

Example Solve $x^3 = 27$.

$$x = \sqrt[3]{27}$$ ← Take the cube root of both sides.

$$x = 3$$ ← 27 INV y^x 3 =

PRACTICE

Solve.

1. $x^2 = 49$ **2.** $x^2 = 19$ **3.** $x^3 = 8$ **4.** $x^3 = 64$

5. $x^3 = -125$ **6.** $x^2 + 25 = 36$ **7.** $x^2 - 16 = 81$ **8.** $49 - x^2 = 25$

Skill 17 Solving Proportions

You can solve some proportions by "undoing." If the variable is in the numerator of one of the equal fractions, you can use multiplication to undo the division.

Example Solve the proportion $\frac{x}{14} = \frac{8}{35}$.

$$\frac{x}{14} = \frac{8}{35}$$

$$14 \cdot \frac{x}{14} = 14 \cdot \frac{8}{35}$$ ← Multiply both sides by 14.

$$x = \frac{16}{5}$$ ← Simplify.

$$x = 3.2$$

Name ______________________________ Date ______________

Skill 17 *(continued)*

You can solve any proportion by using the fact that the two cross products are equal.

Proportion: $\frac{a}{b} = \frac{c}{d}$ *Cross products:* $ad = bc$

Example Solve the proportion $\frac{15}{x} = \frac{20}{21}$.

$$\frac{15}{x} = \frac{20}{21}$$

$15 \cdot 21 = x \cdot 20$ ← Set the two cross products equal.

$315 = 20x$ ← Simplify.

$\frac{315}{20} = \frac{20x}{20}$ ← Divide both sides by 20.

$15.75 = x$ ← Simplify.

PRACTICE

Solve each proportion.

1. $\frac{x}{12} = \frac{13}{10}$
2. $\frac{18}{25} = \frac{z}{15}$
3. $\frac{y}{8} = \frac{14}{20}$
4. $\frac{24}{x} = \frac{15}{7}$
5. $\frac{28}{w} = \frac{16}{9}$
6. $\frac{11}{8} = \frac{5}{y}$
7. $\frac{17}{40} = \frac{q}{24}$
8. $\frac{12}{19} = \frac{33}{n}$

Skill 18 Using the Zero-Product Property

Zero-Product Property If a product of factors is zero, one or more of the factors must be zero.

Example Solve the equation $(20 - x)(30 - 5x) = 0$.

$20 - x = 0$	or	$30 - 5x = 0$	← At least one of the two factors must be 0.
$20 - x + x = 0 + x$		$30 - 5x + 5x = 0 + 5x$	
$20 = x$		$30 = 5x$	
		$\frac{30}{5} = \frac{5x}{5}$	
The solutions are 20 and 6.		$6 = x$	

PRACTICE

Solve each equation.

1. $x(x + 10) = 0$
2. $2x(4 - x) = 0$
3. $-4x(x - 3) = 0$
4. $(x + 1)(x + 6) = 0$
5. $(x + 4)(x - 7) = 0$
6. $(2x - 1)(2x + 6) = 0$

Name ______________________________ Date ____________

Skill 19 Recognizing Functions

A **function** is a relationship in which each value of the **control variable** determines exactly one value of the **dependent variable.**

Vertical Line Test: A graph with the control variable on the horizontal axis represents a function if no vertical line meets the graph in more than one place.

Example To tell whether each graph is a function, imagine drawing vertical lines through it.

No two points on the graph lie on the same vertical line.

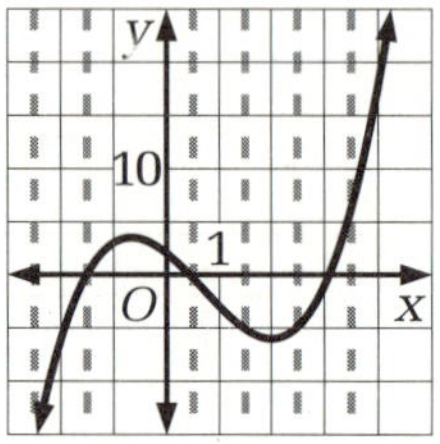

The graph represents a function.

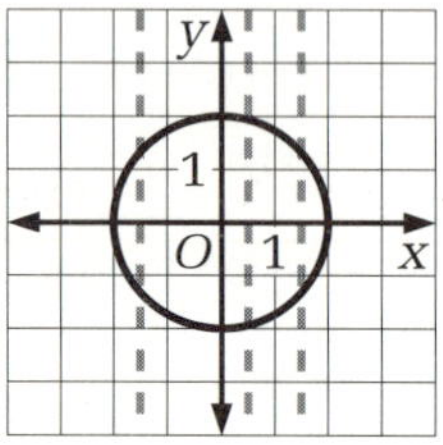

Two points lie on the same vertical line.

The graph does not represent a function.

PRACTICE

Tell whether each graph represents a function when x is the control variable.

1.

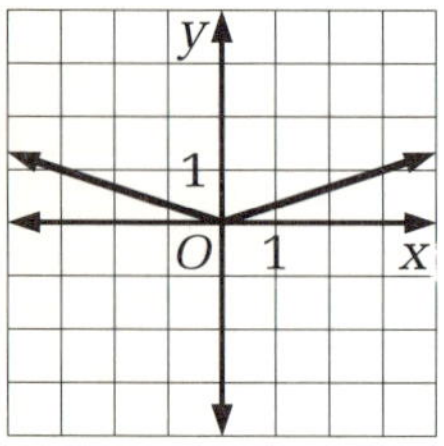

2.

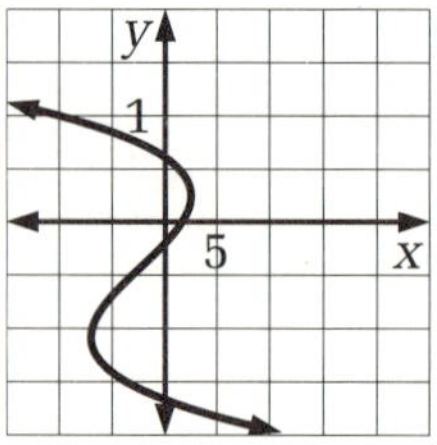

3.

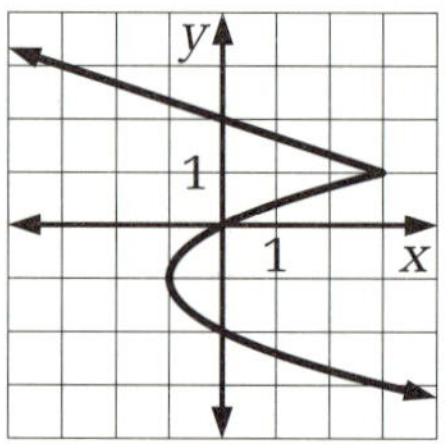

4.

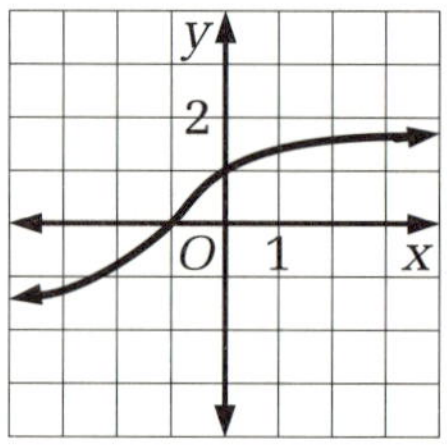

Skill 20 Making a Table of Values to Graph

Example Make a table of values and graph the function $y = 2x + 1$.

$y = 2x + 1$		
x	y	(x, y)
−2	−3	(−2, −3)
−1	−1	(−1, −1)
0	1	(0, 1)
1	3	(1, 3)

Find coordinates of several points. Some easy ones to find are when x is −1, 0, 1.

Plot the ordered pairs. Connect them.

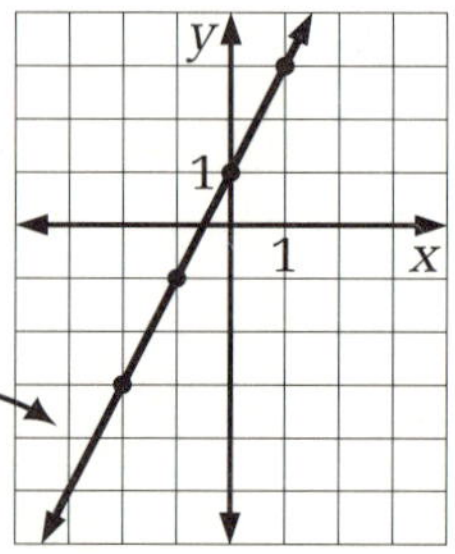

Name ______________________ Date ______________

Skill 20 *(continued)*

Example Make a table of values and graph the function $y = x^2 - 3$.

$y = x^2 - 3$		
x	**y**	**(x, y)**
−3	6	(−3, 6)
−1	−2	(−1, −2)
0	−3	(0, −3)
1	−2	(1, −2)
3	6	(3, 6)

Choose enough values to see the shape of the graph.

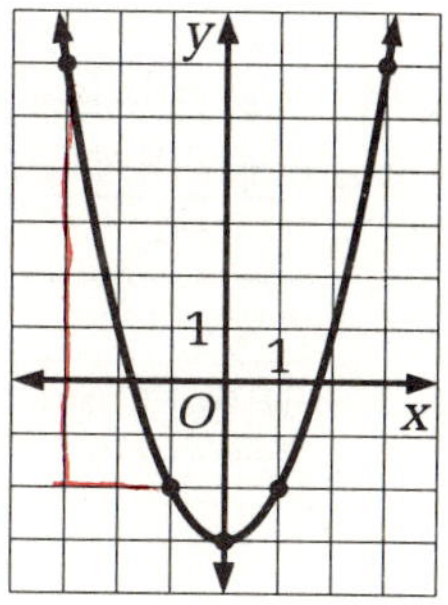

PRACTICE

Make a table of values for each function and graph the function.

1. $y = x - 3$

2. $y = 3x - 4$

3. $y = -2x + 3$

4. $y = \frac{1}{2}x + 1$

5. $y = |x|$

6. $y = x^2 + 1$

7. $y = 2x^2 - 5$

8. $y = \frac{1}{x}$

Skill 21 Finding Slope

The **slope** of a line is the ratio of *rise* to *run*.

Example Find the slope of $\overline{AB}$.

$$\text{slope} = \frac{\text{rise}}{\text{run}} = \frac{4.8}{9.6} = 0.5$$

The slope is 0.5.

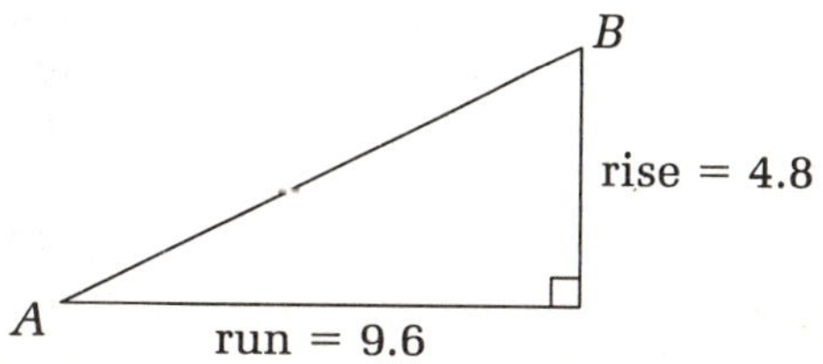

You can find the slope of a line in a coordinate system if you know two points on the line.

Example You can find the slope of the line shown using the points (−2, 1) and (3, 4).

vertical change → Subtract the *y*-coordinates.

$$\text{slope} = \frac{\text{rise}}{\text{run}} = \frac{4 - 1}{3 - (-2)} = \frac{3}{5} = 0.6$$

horizontal change → Subtract the *x*-coordinates in the same order.

The slope of the line is 0.6.

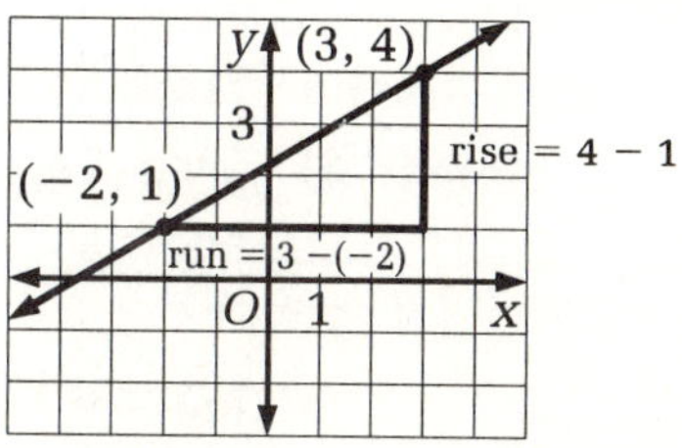

Skill 21 *(continued)*

The **slope of a horizontal line** is 0. The **slope of a vertical line** is undefined. A **negative slope** indicates that the line is "downhill" from left to right.

Example Find the slope of the line $y = 3$.

Two points on this line are (0, 3) and (4, 3).

slope $= \frac{3-3}{4-0} = \frac{0}{4}$, or 0

The slope is 0. (The line is horizontal.)

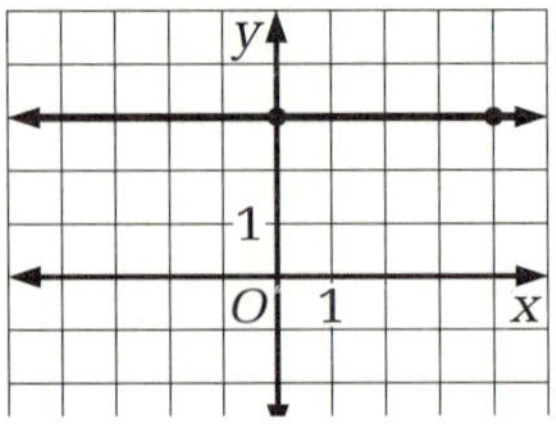

PRACTICE

Find the slope of each line.

1. 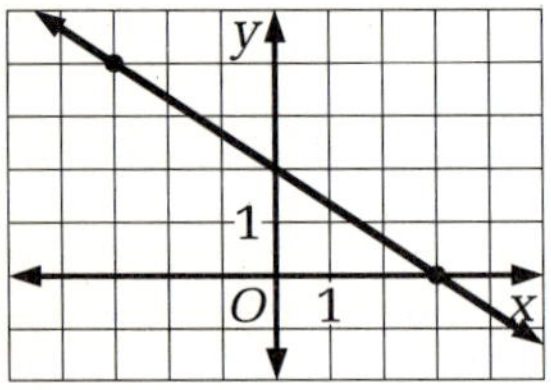

2. the line through $(1, -4)$ and $(3, 7)$
3. the line through $(-6, 2)$ and $(4, -2)$
4. the line through $(5, 0)$ and $(5, 5)$

Skill 22 Modeling Direct Variation

When the ratio of two variable quantities is a constant k, their relationship is called a **direct variation.** You can model a direct variation with an equation of the form $\frac{y}{x} = k$, or $y = kx$.

One quantity varies directly with another quantity.

dependent variable → $y = kx$ ← control variable

↑ variation constant

Example The pressure on an underwater object varies directly with the depth of the object below the surface. Suppose the pressure on an object at a depth of 50 ft is 21.7 lbs/in.[2] Find the pressure on an object at a depth of 85 ft.

Method ❶ Use a proportion. The ratio depth : pressure is constant.

depth → / pressure → $\frac{50}{21.7} = \frac{85}{y}$

$50y = (21.7)(85)$ ← Set the cross products equal.

$y = \frac{(21.7)(85)}{50} = 36.89$

Continued on next page.

Name ______________________________ Date ______________

Skill 22 *(continued)*

Method ❷ Use the equation $y = kx$.

Step 1 First find the variation constant.

Let x = the depth (ft) and y = the pressure (lbs/in.^{2}).

$y = kx$ ← Write the general form of the equation.

$21.7 = k \cdot 50$ ← Substitute **50** for ***x*** and **21.7** for ***y***.

$k = \frac{21.7}{50} = 0.434$ ← Solve for *k*, the variation constant.

Step 2 $y = kx$ ← Write the general form of the equation.

$y = (0.434)(85)$ ← Substitute 0.434 for *k* and 85 for *x*.

$y = 36.89$

The pressure is 36.89 lbs/in.^{2}

PRACTICE

1. The distance a car travels at a constant speed varies directly with time. Suppose the car goes 92 miles in 2 h. How far would it go in 3.5 h?
2. The bounce height of a rubber ball varies directly with the drop height. A ball dropped from a height of 6.5 ft bounces to a height of 5.2 ft. How high will the ball bounce if dropped from 5.5 ft?

Skill 23 Using $y = mx + b$

One form of the equation of a line is **slope-intercept form:**

$y = mx + b$

slope → m; vertical intercept → b

Example Find the slope and the vertical intercept of the line $y = \frac{2}{3}x - 5$.

The line is in the form $y = mx + b$. Therefore, the slope is $\frac{2}{3}$, and the vertical intercept is -5.

Example Write an equation for the line shown at the right.

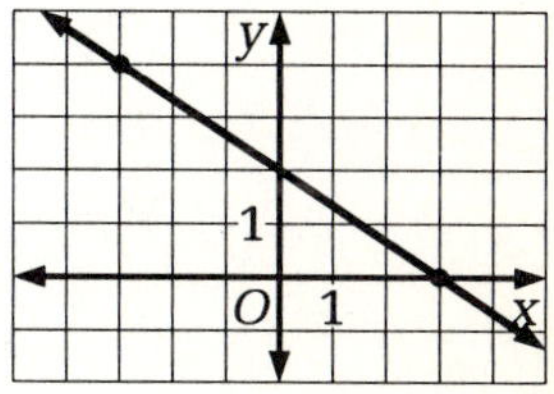

Step 1 Use $(-3, 4)$ and $(3, 0)$ to find the slope m.

$$m = \frac{0-4}{3-(-3)} = \frac{-4}{6} = -\frac{2}{3}$$

Step 2 The graph crosses the vertical axis at $(0, 2)$, so $b = 2$.

Step 3 Substitute the values in $y = mx + b$: $y = -\frac{2}{3}x + 2$.

Name ______________________ Date ____________

Skill 23 *(continued)*

PRACTICE

Find the slope and the vertical intercept of each line.

1. $y = 3x - 4$ **2.** $y = -2.5x + 11.2$ **3.** $y = 6 - 7x$ **4.** $y = 4x$

Write an equation of each line in the form $y = mx + b$.

5.

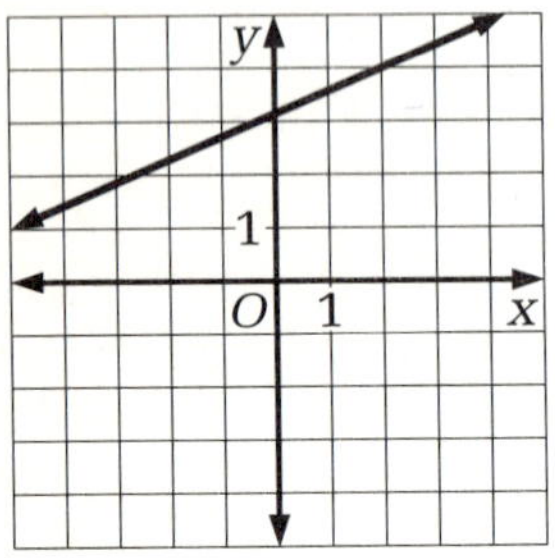

6.

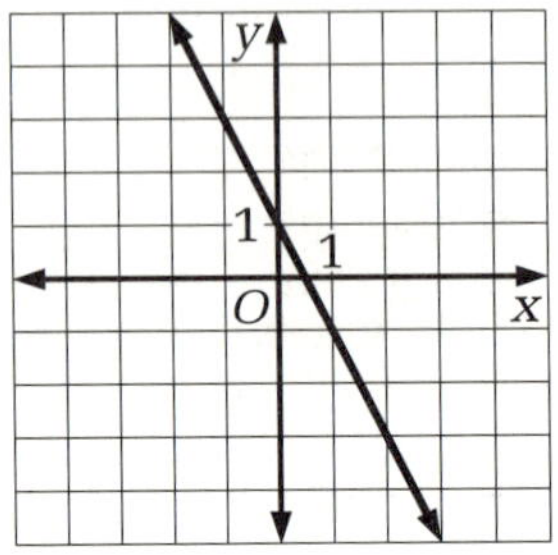

7.

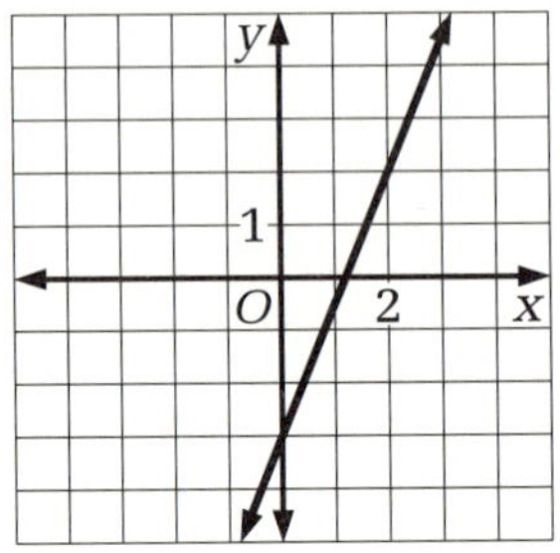

Skill 24 Using Intercepts to Graph

The **horizontal intercept** of a graph is the x-coordinate of the point where the graph crosses the x-axis. The **vertical intercept** is the y-coordinate of the point where the graph crosses the y-axis.

The graph at the right crosses the horizontal axis at (3, 0) and the vertical axis at (0, 2).

The horizontal intercept of the graph is 3.

The vertical intercept of the graph is 2.

If you know the equation of a line, one way to graph the line is to use the intercepts.

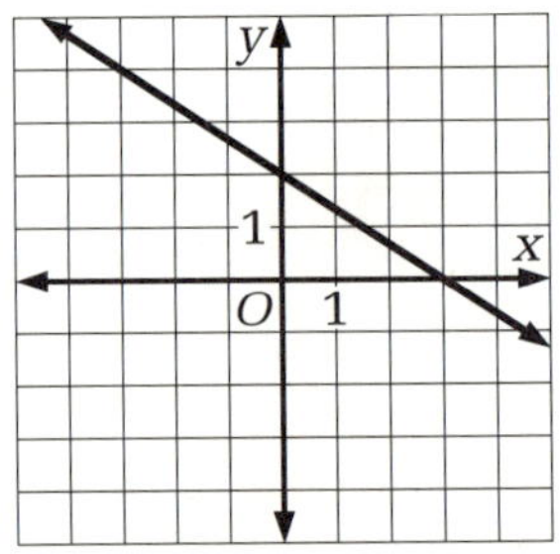

Example Use the intercepts to graph the line $2x - 5y = 15$.

To find the vertical intercept, substitute 0 for x:

$$2x - 5y = 15$$
$$2(0) - 5y = 15$$
$$-5y = 15$$
$$y = -3$$

To find the horizontal intercept, substitute 0 for y:

$$2x - 5y = 15$$
$$2x - 5(0) = 15$$
$$2x = 15$$
$$x = 7.5$$

Plot the points $(0, -3)$ and $(7.5, 0)$.
Connect the points.

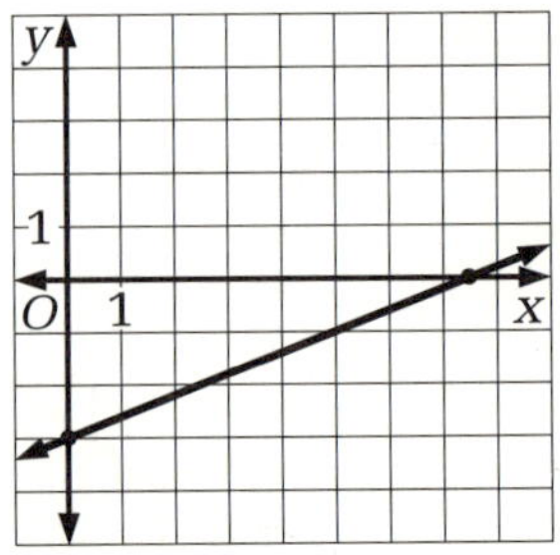

Name ______________________________ Date ______________

Skill 24 *(continued)*

PRACTICE

Use the vertical and horizontal intercepts to graph each equation.

1. $3x + y = 6$
2. $5x - 2y = 10$
3. $-3x + 3y = 9$
4. $4x = y + 8$
5. $y = 2x - 3$
6. $y = -2.5x + 5$

Skill 25 Graphing an Inequality in Two Variables

To graph a **linear inequality** in two variables, such as $y > x + 2$, you shade a region on a coordinate plane whose edge is a line. The edge is called a **boundary line.** The shaded part of the graph of a linear inequality is the **solution region.**

Example Graph the inequality $y > x + 2$.

First graph the boundary line: $y = x + 2$. Since the inequality symbol is ">," you use a dashed line for the boundary line. Each point on the boundary line has y-coordinate $x + 2$. Therefore, the points for which $y > x + 2$ will lie *above* the boundary line. Shade this region.

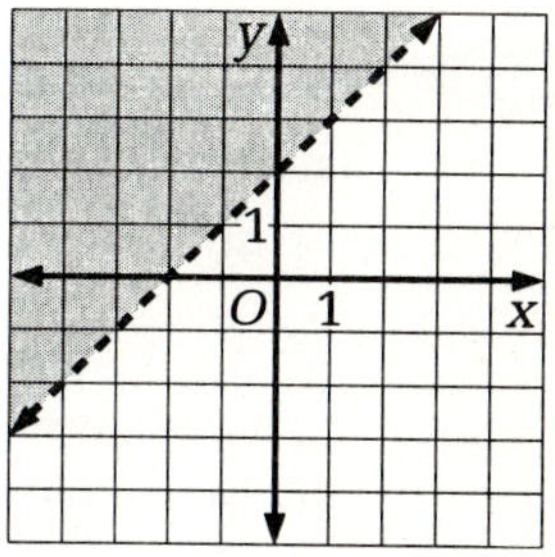

Example Graph the inequality $x + 2y \le 6$.

First rewrite the inequality in slope-intercept form.

$$x + 2y \le 6$$

$$x + 2y - x \le 6 - x$$ ← Subtract x from both sides.

$$2y \le 6 - x$$

$$\frac{1}{2} \cdot 2y \le \frac{1}{2}(6 - x)$$ ← Multiply both sides by $\frac{1}{2}$.

$$y \le 3 - \frac{1}{2}x$$

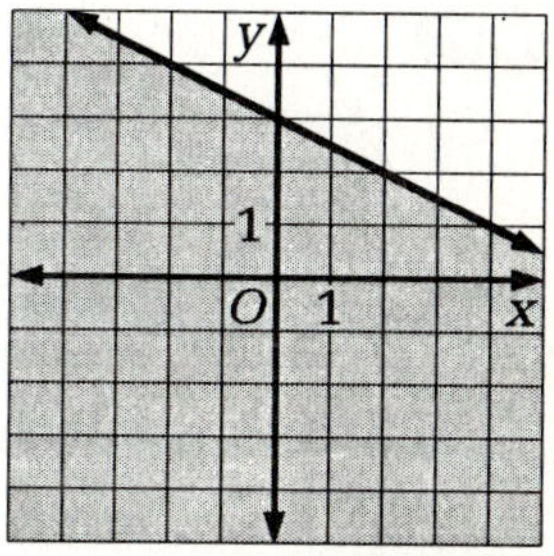

Graph the boundary line $y = 3 - \frac{1}{2}x$.

Since the symbol in the inequality is "≤," the boundary line is a solid line.

Each point on the boundary line has y-coordinate $3 - \frac{1}{2}x$. Therefore, the solution region lies *below* and includes the boundary line. Shade this region.

PRACTICE

Graph each inequality.

1. $y > -x + 3$
2. $y \le 2x - 4$
3. $y < \frac{1}{2}x + 1$
4. $2x + 3y \ge 6$
5. $x - 3y > 3$
6. $3x + 2y \le 9$

Name ______________________ Date ____________

Skill 26 Finding Unknown Angle Measures

An angle whose measure is 90° is called a **right angle.** An angle whose measure is 180° is called a **straight angle.**

Two angles whose measures add up to 90° are **complementary.**
Two angles whose measures add up to 180° are **supplementary.**

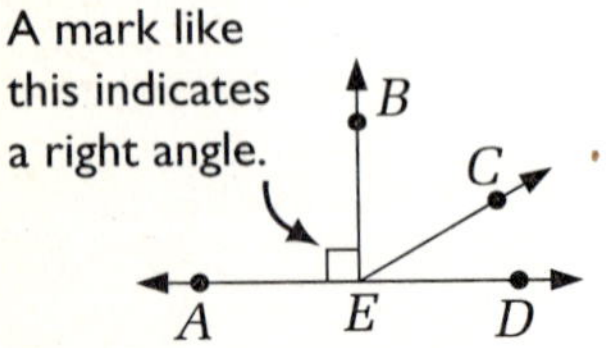

∠*AEB* is a right angle.
∠*BEC* and ∠*CED* are complementary.

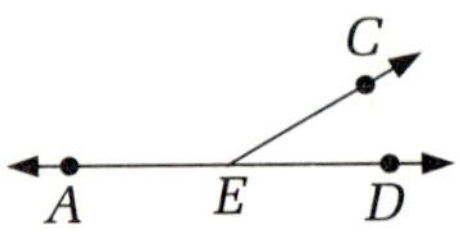

∠*AED* is a straight angle.
∠*AEC* and ∠*CED* are supplementary.

Example Find the unknown angle measure in the figure.

The sum of the measures of the angles of a triangle is 180°.

$x + 40 + 30 = 180$

$x = 180 - 70$

$= 110$

The unknown angle measure is 110°.

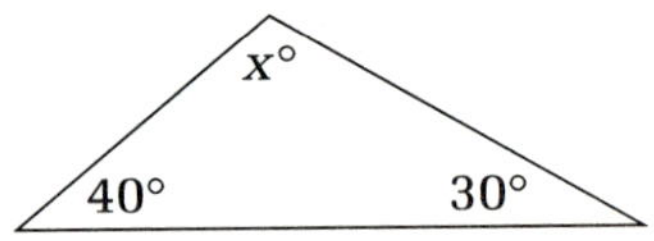

Example Find the unknown angle measures in the figure.

Vertical angles are equal in measure.

$\angle AOD$ and $\angle BOC$ are vertical angles, so $y = 55$.

$\angle AOB$ and $\angle BOC$ are supplementary, so
$x = 180 - 55 = 125$.

The angle measures are 125° and 55°.

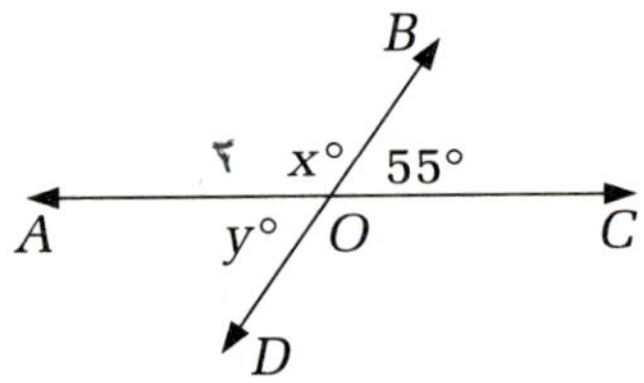

PRACTICE

Find each unknown angle measure.

1.

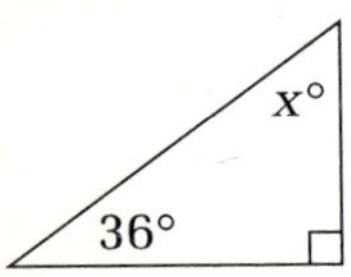

2.

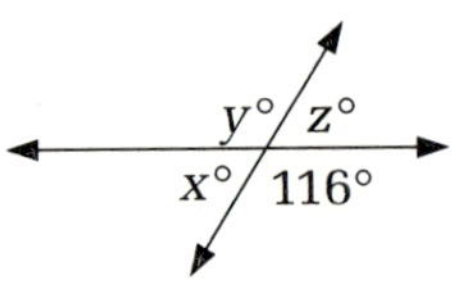

3.

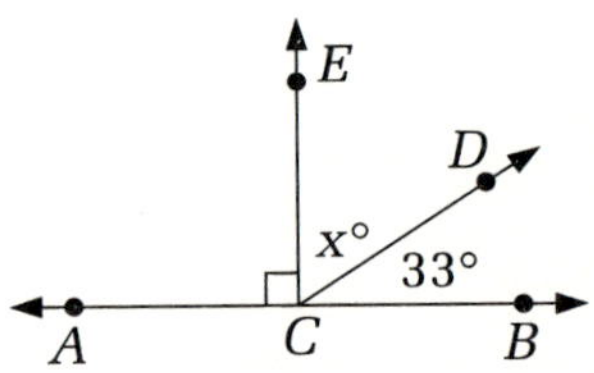

Name ______________________________ Date ______________

Skill 27 Reading Diagrams

Certain marks in a geometry diagram indicate special relationships between lines, segments, and angles.

A small square indicates a right angle.

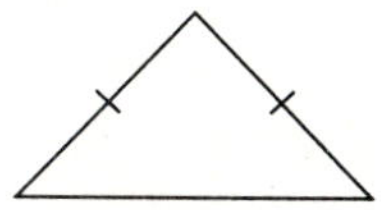

Tick marks indicate sides that are equal in measure.

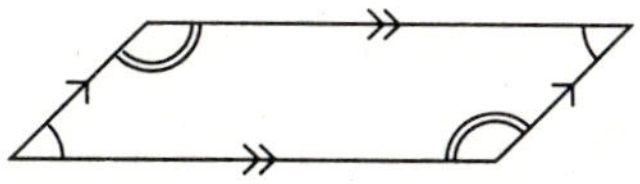

Arcs indicate angles that are equal in measure.

Arrowheads indicate parallel lines or segments.

Example List all the relationships indicated by marks in the figure.

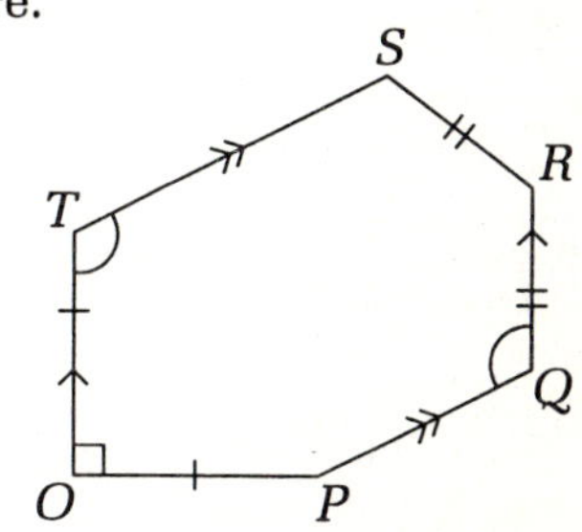

Angles Since $\angle TOP$ is marked with a small square, $\angle TOP$ is a right angle.

Since $\angle T$ and $\angle Q$ both have one arc, $\angle T$ and $\angle Q$ are equal in measure.

Sides Since sides TO and OP both have one tick mark, side $TO \cong$ side OP.

Since sides SR and RQ both have two tick marks, side $SR \cong$ side RQ.

Since sides RQ and TO both have one arrowhead, side RQ is parallel to side TO.

Since sides PQ and ST both have two arrowheads, side PQ is parallel to side ST.

PRACTICE

List all the relationships indicated by marks in each figure.

1.

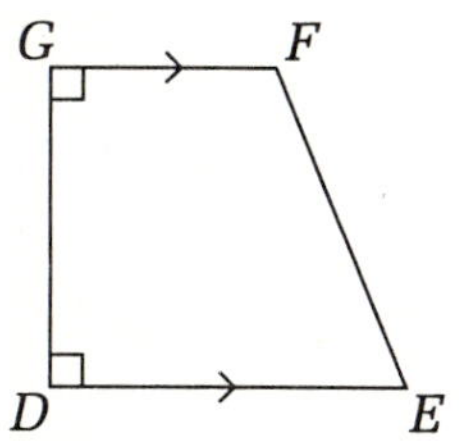

2.

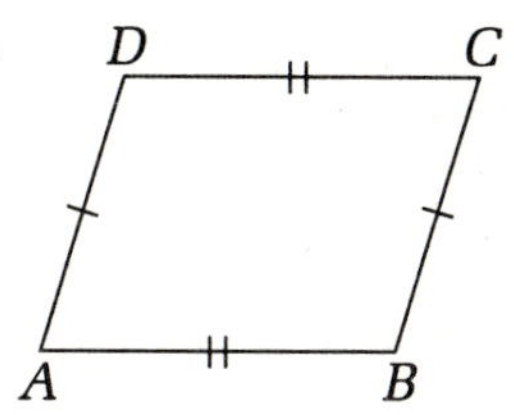

3.

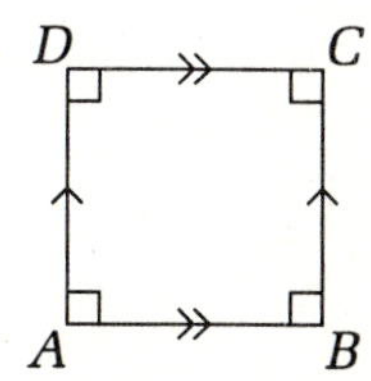

4.

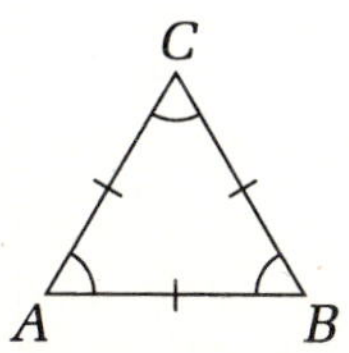

5.

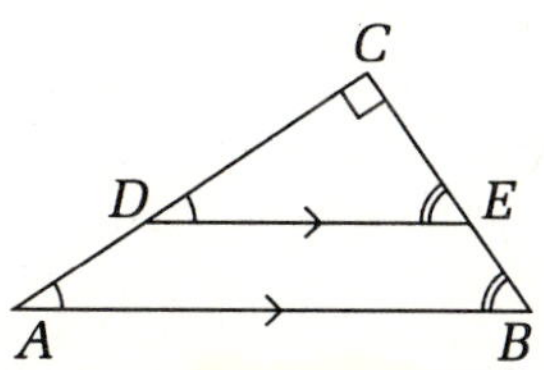

6.

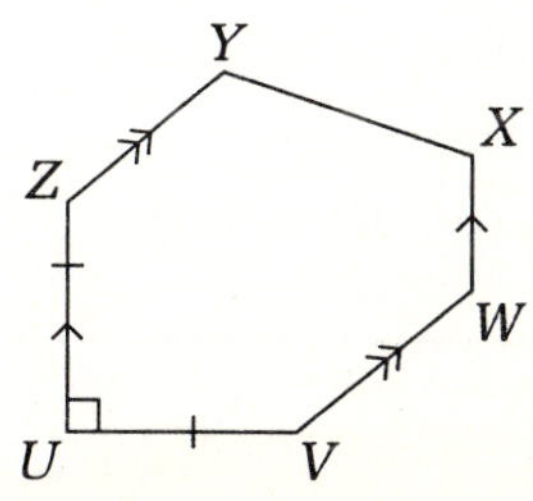

Name ______________________ Date ______________

Skill 28 Using Formulas from Geometry

In formulas from geometry:

A = area S.A. = surface area V = volume

Parallelogram

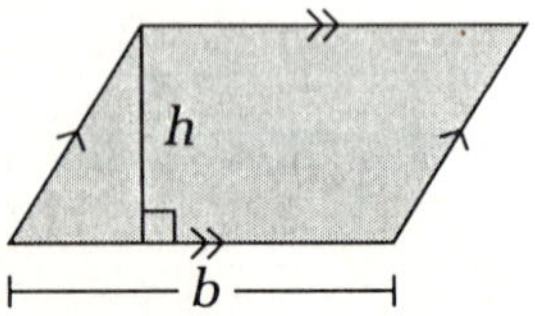

A = base × height

$A = bh$

Triangle

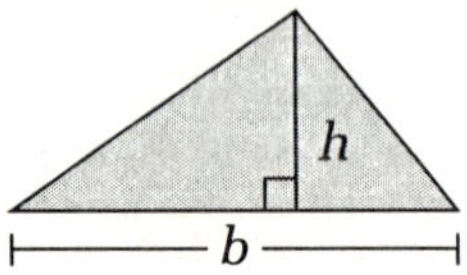

$A = \frac{1}{2} \times$ base × height

$A = \frac{1}{2}bh$

Trapezoid

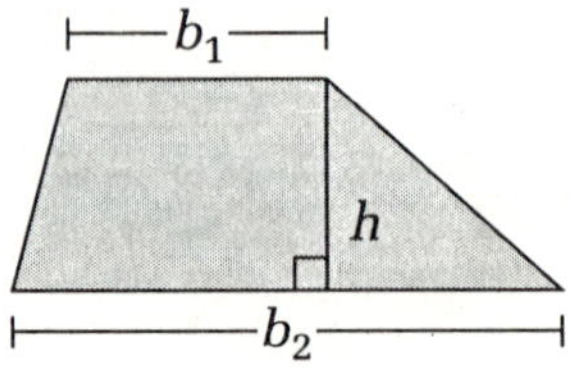

$A = \frac{1}{2} \times$ sum of bases × height

$A = \frac{1}{2}(b_1 + b_2)h$

Circle

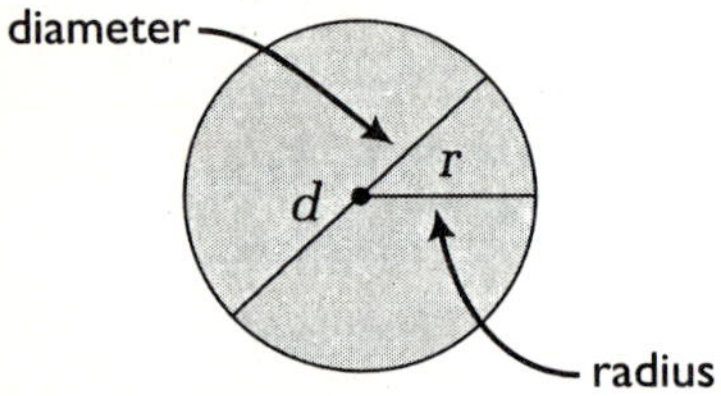

Circumference of a circle:

$C = 2\pi r$ $C = \pi d$

$A = \pi r^2$

Sector

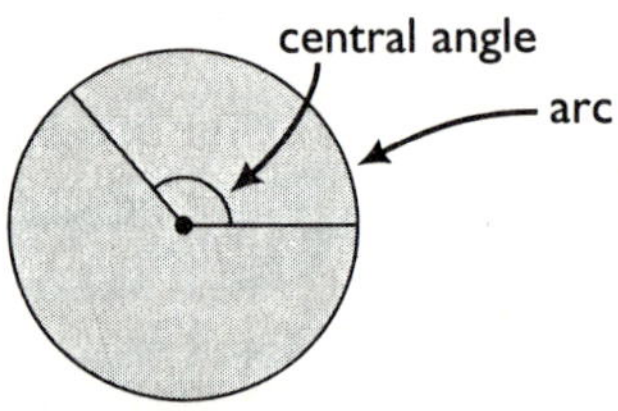

$\frac{\text{arc length}}{\text{circumference}} = \frac{\text{central } \angle}{360°}$

$\frac{\text{area of sector}}{\text{area of circle}} = \frac{\text{central } \angle}{360°}$

Prism

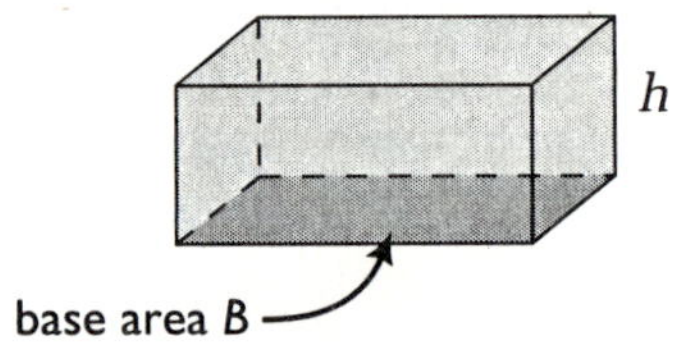

S.A. = areas of two bases + areas of faces

V = area of base × height of prism

Cylinder

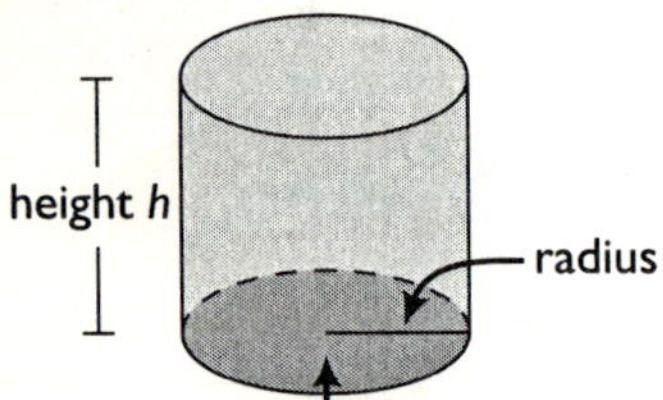

base area $B = \pi r^2$

S.A. = areas of two bases + area of curved surface

S.A. $= 2\pi r^2 + 2\pi rh$

V = area of base × height

$V = Bh = \pi r^2 h$

Pyramid

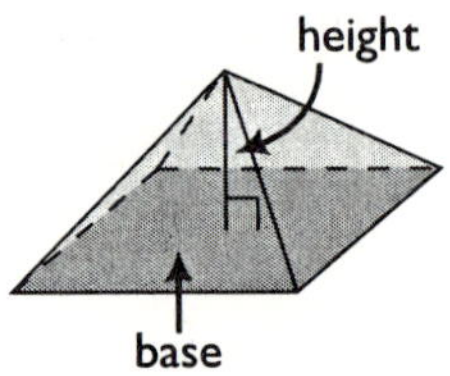

S.A. = area of base + areas of faces

$V = \frac{1}{3} \times$ area of base × height

$V = \frac{1}{3}Bh$

Cone

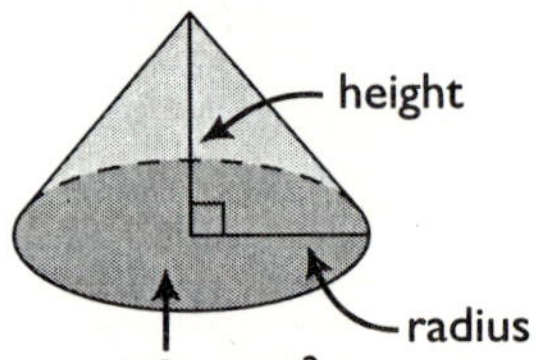

base area $B = \pi r^2$

$V = \frac{1}{3} \times$ area of base × height

$V = \frac{1}{3}Bh = \frac{1}{3}\pi r^2 h$

Name ______________________ Date ____________

Skill 28 *(continued)*

Example Find the area of the trapezoid shown.

$A = \frac{1}{2}(b_1 + b_2)h$ ← Write the formula for the area of a trapezoid.

$A = \frac{1}{2}(21 + 33)13$ ← Substitute **21** for b_1, **33** for b_2, and **13** for h.

$= \frac{1}{2}(54)13 = 351$

The area is 351 cm^2. ← cm × cm = cm^2

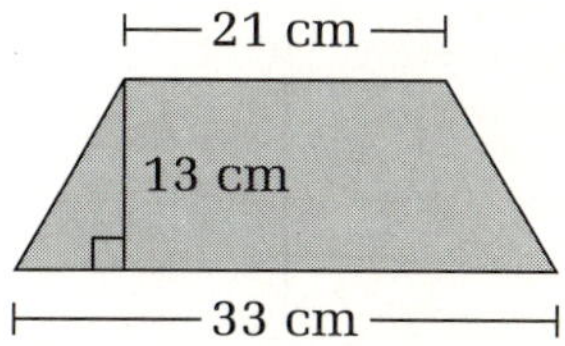

Example Use the the triangular prism shown.

Find: **a.** the volume **b.** the surface area

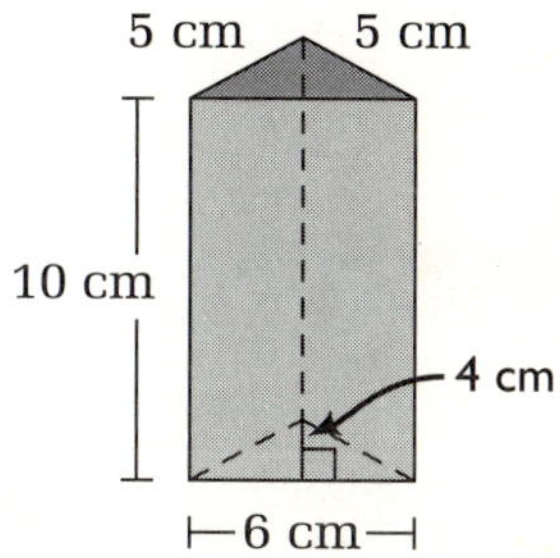

a. Volume of a prism = area of base × height of prism.

Step 1 Find the area of the triangular base of the prism.

$B = \frac{1}{2}bh$ ← Write the formula for the area of a triangle.

$= \frac{1}{2}(6)(4)$ ← Substitute **6** for ***b*** and **4** for ***h***.

$= 12$

Step 2 Find the volume of the prism.

$V = Bh$ ← Write the formula for the volume of a prism.

$= 12 \cdot 10$ ← Substitute **12** for ***B*** and **10** for ***h***.

$= 120$

The volume of the prism is 120 cm^3. ← cm × cm × cm = cm^3

b. Surface area of a prism = area of two bases + area of three faces

$$\text{S.A.} = \overbrace{12 + 12}^{} + \overbrace{6 \cdot 10 + 5 \cdot 10 + 5 \cdot 10}^{}$$

$$= 184$$

The surface area is 184 cm^2.

PRACTICE

Find the area of each figure.

1. a parallelogram of height 10 ft with a base of length 16 ft
2. a trapezoid of height 3 in. with bases of length 5 in. and 8 in.
3. a sector with central angle 40° and radius 9 in.

Find the surface area of each space figure.

4. a rectangular prism with dimensions 18 in., 12 in., and 4.5 in.
5. a cylinder of height 20 cm and with a base of radius 5 cm

Find the volume of each space figure.

6. a triangular prism whose height is 8 cm and whose base area is 25 cm^2
7. a cylinder of height 8.5 m and with a base of radius 2 m
8. a pyramid of height 12 in., whose base is a square with sides of length 10 in.

Name ______________________ Date __________

Skill 29 Finding Sine, Cosine, and Tangent Ratios

Suppose $\angle A$ is one of the acute angles of a right triangle. The **trigonometric ratios** for $\angle A$ are ratios of the measures of the sides of the triangle.

Sine of $\angle A$ $\quad \sin A = \dfrac{\text{leg opposite } \angle A}{\text{hypotenuse}} = \dfrac{BC}{AB}$

Cosine of $\angle A$ $\quad \cos A = \dfrac{\text{leg adjacent to } \angle A}{\text{hypotenuse}} = \dfrac{AC}{AB}$

Tangent of $\angle A$ $\quad \tan A = \dfrac{\text{leg opposite } \angle A}{\text{leg adjacent to } \angle A} = \dfrac{BC}{AC}$

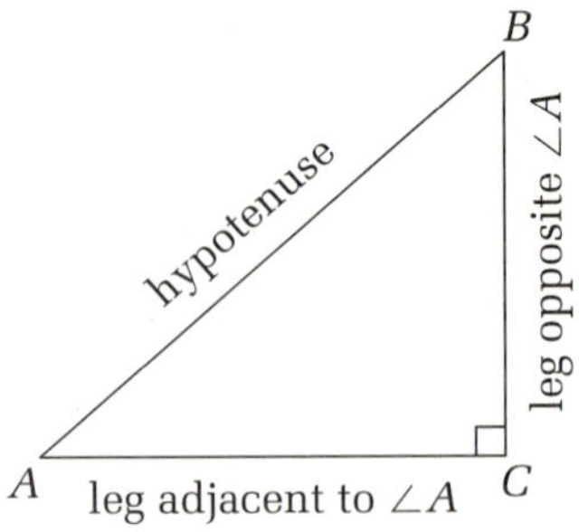

Example Write each trigonometric ratio as a fraction and as a decimal rounded to the nearest hundredth.

a. $\sin P$ **b.** $\cos P$ **c.** $\tan P$

d. $\sin Q$ **e.** $\cos Q$ **f.** $\tan Q$

a. $\sin P = \dfrac{\text{leg opposite } \angle P}{\text{hypotenuse}} = \dfrac{15}{17} \approx 0.88$

b. $\cos P = \dfrac{\text{leg adjacent to } \angle P}{\text{hypotenuse}} = \dfrac{8}{17} \approx 0.47$

c. $\tan P = \dfrac{\text{leg opposite } \angle P}{\text{leg adjacent to } \angle P} = \dfrac{15}{8} \approx 1.88$

d. $\sin Q = \dfrac{\text{leg opposite } \angle Q}{\text{hypotenuse}} = \dfrac{8}{17} \approx 0.47$

e. $\cos Q = \dfrac{\text{leg adjacent to } \angle Q}{\text{hypotenuse}} = \dfrac{15}{17} \approx 0.88$

f. $\tan Q = \dfrac{\text{leg opposite } \angle Q}{\text{leg adjacent to } \angle Q} = \dfrac{8}{15} \approx 0.53$

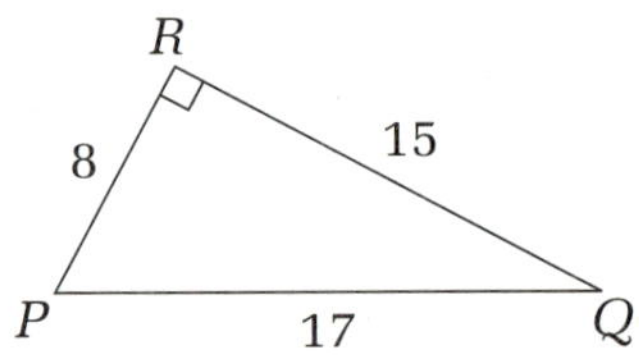

PRACTICE

Use the triangle at the right. Find the value of each trigonometric ratio.

1. $\sin A$ **2.** $\cos A$ **3.** $\tan A$

4. $\sin B$ **5.** $\cos B$ **6.** $\tan B$

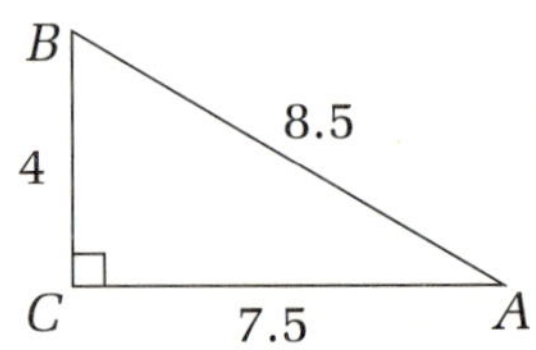

Draw and label a right triangle that fits each description.

7. $\triangle ABC$ with $\sin A = \dfrac{\text{leg opposite } \angle A}{\text{hypotenuse}} = \dfrac{3}{5}$

8. $\triangle XYZ$ with $\cos X = \dfrac{\text{leg adjacent to } \angle X}{\text{hypotenuse}} = \dfrac{5}{13}$

Name ______________________________ Date ______________

Skill 30 Finding Missing Lengths in Right Triangles

You can use the trigonometric ratios to find missing lengths in right triangles. The ratios are programmed into scientific calculators.

Example Find the missing length in the triangle at the right.

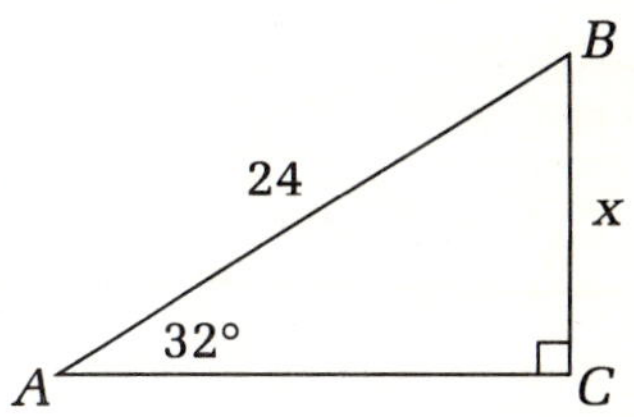

Since x is the measure of the leg opposite $\angle A$, and 24 is the measure of the hypotenuse of the right triangle, you use the sine ratio.

$$\sin \angle A = \frac{\text{length of leg opposite } \angle A}{\text{length of hypotenuse}}$$

$$\sin 32° = \frac{x}{24}$$

$$0.5299 \approx \frac{x}{24}$$ ← Make sure your calculator is in degree mode. Use the **sin** key.

$$0.5299 \cdot 24 \approx \frac{x}{24} \cdot 24$$ ← Multiply both sides by 24.

$$12.7 \approx x$$

Example Find the missing length in the triangle at the right.

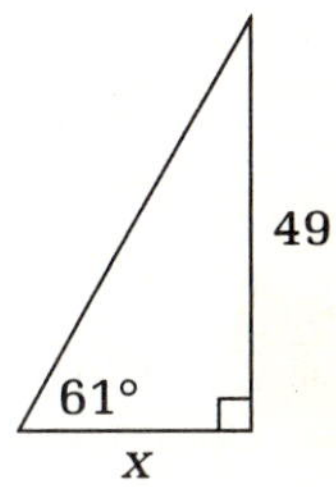

Use the tangent ratio.

$$\tan 61° = \frac{49}{x}$$

$$1.804 \approx \frac{49}{x}$$ ← Use the **tan** key on your calculator.

$$1.804 \cdot x \approx \frac{49}{x} \cdot x$$ ← Multiply both sides by x.

$$1.804x \approx 49$$

$$\frac{1.804x}{1.804} \approx \frac{49}{1.804}$$ ← Divide both sides by 1.804.

$$x \approx \frac{49}{1.804} \approx 27.2$$

PRACTICE

Find each missing length.

1.

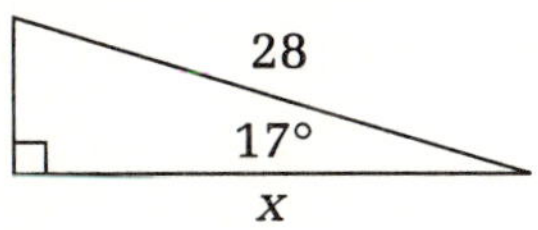

2.

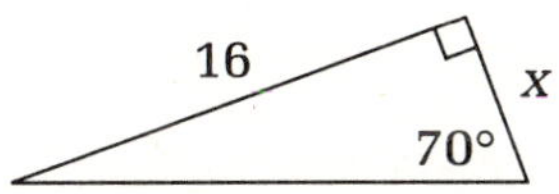

3.

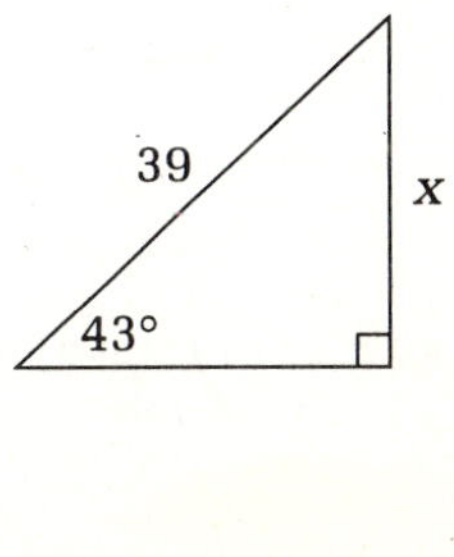

Name ______________________________ Date ______________

Skill 31 Finding Missing Lengths

The Pythagorean Theorem In any right triangle with legs of length a and b and hypotenuse of length c,

$$a^2 + b^2 = c^2.$$

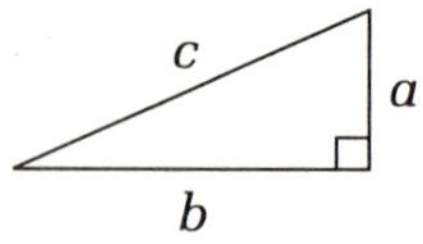

Example Find the missing length in the triangle at the right.

$c^2 = 6.4^2 + 12^2$ ← Use the Pythagorean theorem: $c^2 = a^2 + b^2$

$c^2 = 40.96 + 144$

$c^2 = 184.96$

$c = \sqrt{184.96}$ ← Find the positive square root.

$c = 13.6$

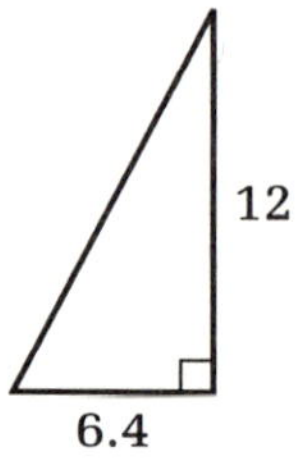

PRACTICE

Find the missing length in the right triangle with legs *a* and *b* and hypotenuse *c*.

1. $a = 5$, $b = 12$, $c = ?$ **2.** $a = 10$, $c = 14.5$, $b = ?$ **3.** $b = 24$, $c = 75$, $a = ?$

Skill 32 Determining a Right Triangle

If the lengths of the sides of a triangle satisfy the relationship $a^2 + b^2 = c^2$, the triangle is a right triangle with hypotenuse c.

Example Is the triangle with sides of the given lengths a right triangle?

a. 15, 20, 25 **b.** 8, 10, 13

If the triangle is a right triangle, the longest side must be the hypotenuse.

a.
$a^2 + b^2 \stackrel{?}{=} c^2$

$15^2 + 20^2 \stackrel{?}{=} 25^2$

$225 + 400 \stackrel{?}{=} 625$

$625 = 625$ ✓

The triangle is a right triangle.

b.
$a^2 + b^2 \stackrel{?}{=} c^2$

$8^2 + 10^2 \stackrel{?}{=} 13^2$

$64 + 100 \stackrel{?}{=} 169$

$164 \neq 169$

The triangle is not a right triangle.

PRACTICE

Tell whether a triangle with sides of the given lengths is a right triangle.

1. 6, 8, 10 **2.** 2.1, 2.8, 3.5 **3.** 2.5, 6, 6.5 **4.** 10, 17.5, 20

Name ______________________________ Date ______________

Skill 33 Translating a Figure

A **translation** of a plane figure slides the figure without changing its size or shape and without turning it. Each point of the figure moves the same distance and in the same direction.

Example Translate $\triangle ABC$ 4 units to the right and 1 unit down.

Step 1 Translate each vertex 4 units to the right and 1 unit down:

$A(-4, 4) \rightarrow A'(0, 3)$

$B(-1, 2) \rightarrow B'(3, 1)$

$C(-2, -1) \rightarrow C'(2, -2)$

Each translated vertex is labeled with a prime (′).

Step 2 Connect the vertices. The translated triangle is $\triangle A'B'C'$.

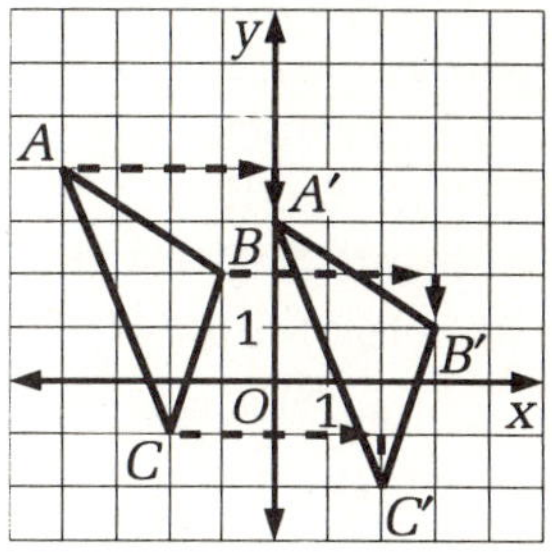

PRACTICE

Draw each translation of $\triangle DEF$.

1. 3 units to the left.
2. 2 units down
3. 2 units to the right and up 3 units
4. 4 units to the left and down 1 unit
5. 1 unit to the right and down 4 units

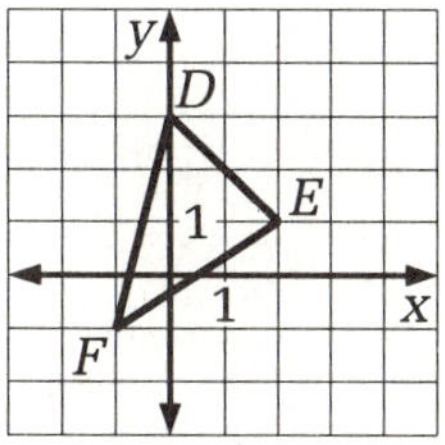

Skill 34 Rotating a Figure

A **rotation** moves each point of a figure through a circular arc about a common point, called the **center of rotation.** Each point of the figure is rotated by the same number of degrees.

Example Rotate $\triangle ABC$ 90° counterclockwise about the origin.

Imagine that $\triangle ABC$ is a paper triangle fixed to a stick, $\overline{OC}$. Suppose you press down on the stick at the center of rotation O and rotate the stick through a quarter of a circle (90°) counterclockwise. The rotated triangle is $\triangle A'B'C'$.

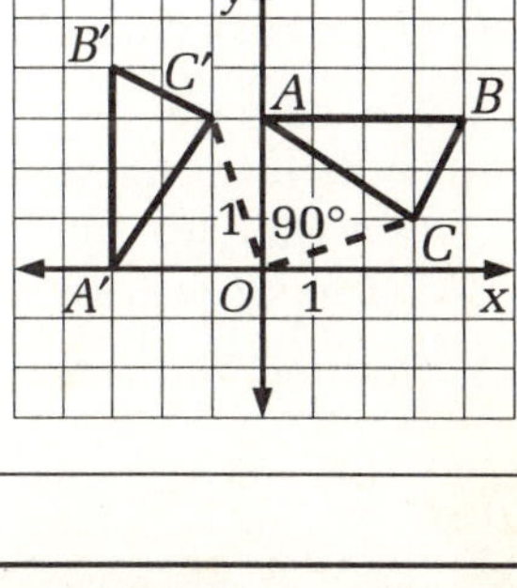

Name ______________________________ Date ______________

Skill 34 *(continued)*

It is easier to draw a rotation on polar graph paper.

Example Rotate $\triangle DEF$ 60° counterclockwise about the origin.

Step 1 Move each vertex along its circle six lines in the counterclockwise direction. (From one line to the next is 10°, and $6 \cdot 10°$ is 60°.) Label the new points D', E', F'.

Step 2 Connect the vertices. The rotated triangle is $\triangle D'E'F'$.

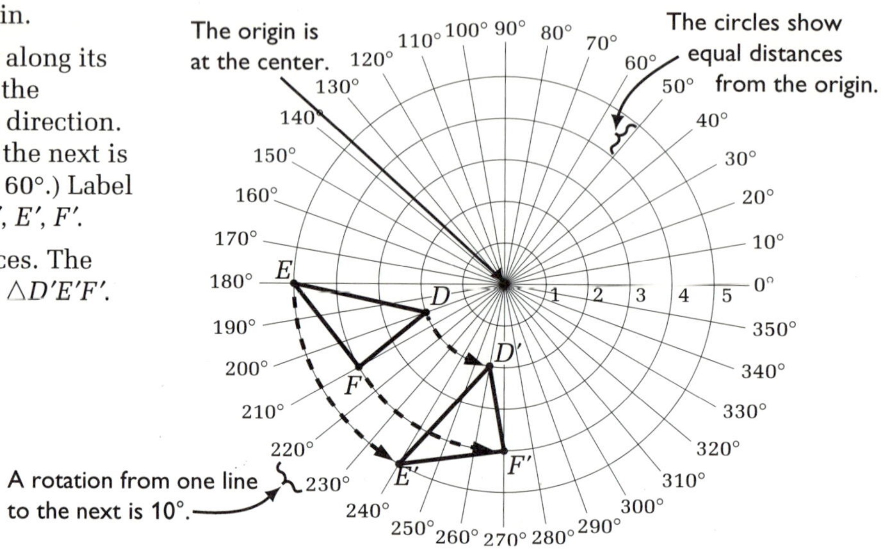

PRACTICE

Draw each rotation of $\triangle PQR$ about the origin.

1. 90° counterclockwise
2. 90° clockwise
3. 180° clockwise
4. 270° counterclockwise

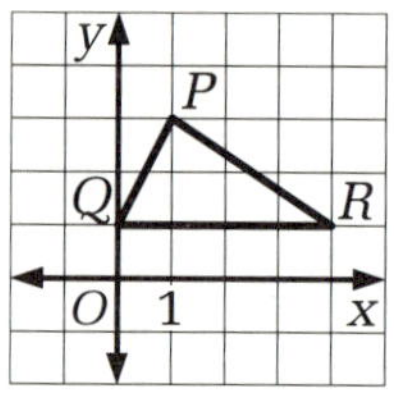

Skill 35 Dilating a Figure

A **dilation** transforms a figure into a similar figure. The ratio of any length in the transformed figure to the corresponding length in the original figure is the **scale factor.**

Example Draw a dilation of $\triangle ABC$, with center P, in which the scale factor is 3.

Step 1 Draw rays from P through the vertices of $\triangle ABC$.

Step 2 Measure the distance from P to each vertex of the original triangle along these rays. Mark points A', B', and C' on these rays that are 3 times as far from P as the corresponding vertex of the original figure.

Step 3 Connect the vertices. The dilated triangle is $\triangle A'B'C'$.

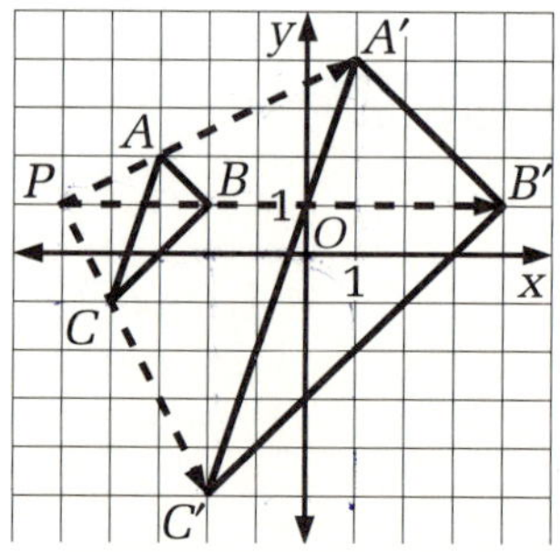

Name ______________________ Date __________

Skill 35 *(continued)*

PRACTICE

Draw a dilation of the parallelogram *JKLM* with each scale factor and with center *J*.

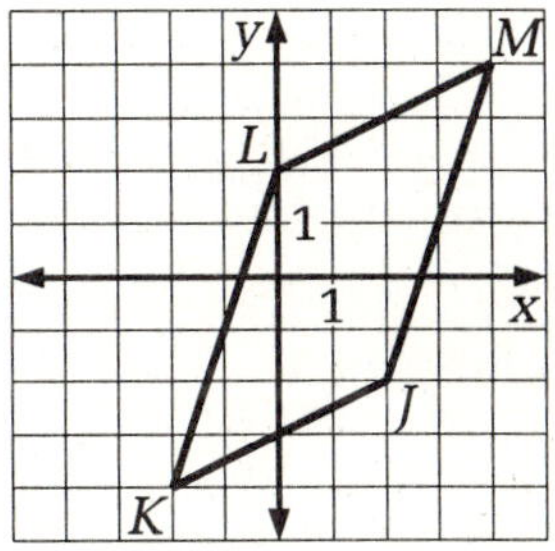

1. scale factor 2
2. scale factor $\frac{1}{2}$
3. scale factor $\frac{3}{2}$
4. scale factor $\frac{5}{2}$

Skill 36 Reflecting a Figure

A **reflection** is a flip over a line. The line is called the **line of reflection.** The reflected figure is congruent to the original figure, but its orientation is reversed.

Example Draw the reflection of $\triangle ABC$ over the x-axis. Label the vertices with primes.

Point A is 3 units above the x-axis.

Point A′ is 3 units below the x-axis.

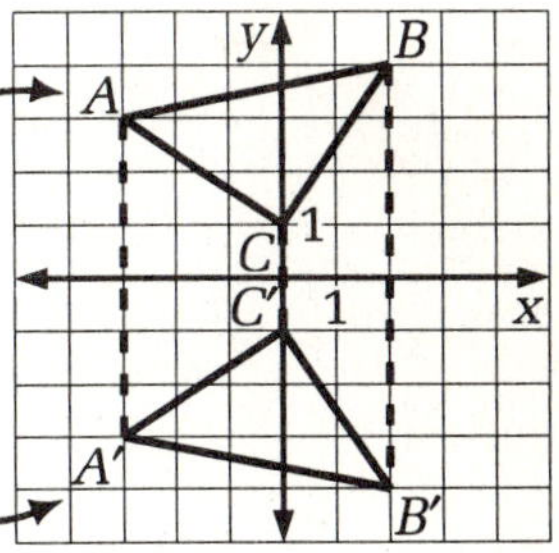

Notice in the sample that a reflection has these properties:

- Each segment connecting a vertex of the original figure with its corresponding vertex in the reflected figure is cut in half by the line of reflection, in this case, the x-axis.
- The vertices of the original figure in *clockwise* order correspond to the vertices of the reflected figure in counterclockwise order.

PRACTICE

Draw a reflection of $\triangle DEF$ over each line.

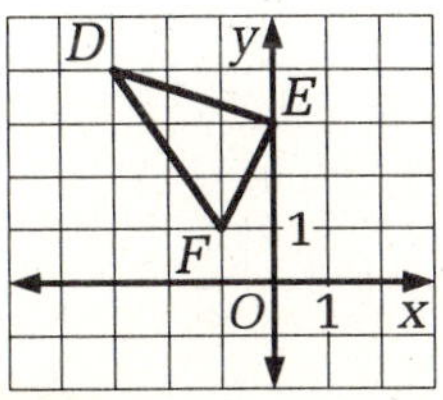

1. the x-axis
2. the y-axis
3. the line $x = 1$
4. the line $y = -1$

Scatter Plots

FOCUS

Make and interpret a scatter plot of data, use scatter plots to make predictions.

The graphing calculators available on the market today allow the user to input a set of ordered pairs and then have the calculator draw a scatter plot of the data points. These calculators can also be used to determine a fitted line for the data which can then be added to the scatter plot.

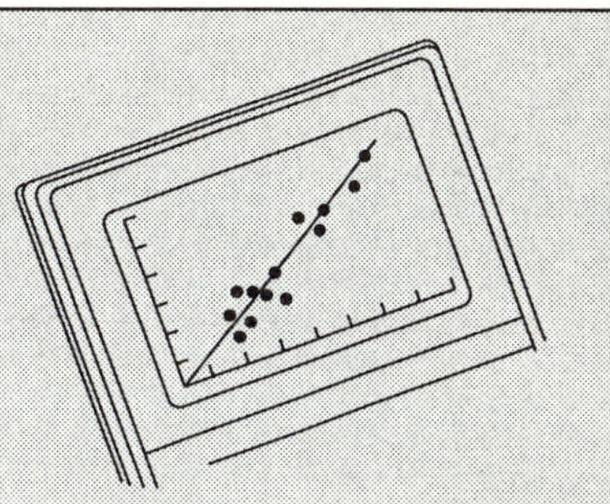

KEY TERMS | **EXAMPLE / ILLUSTRATION**

Scatter plot
a graph that shows the relationship between two data sets

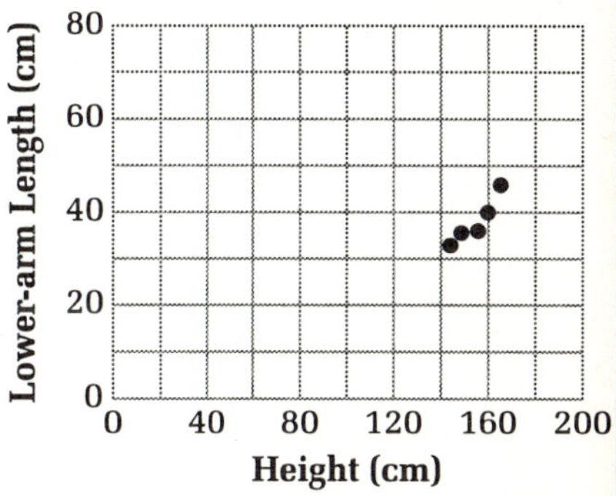

Fitted line
a straight line drawn on a scatter plot that passes close to most of the data points (This line helps to make predictions—the stronger the correlation, the more accurate the predictions will be.)

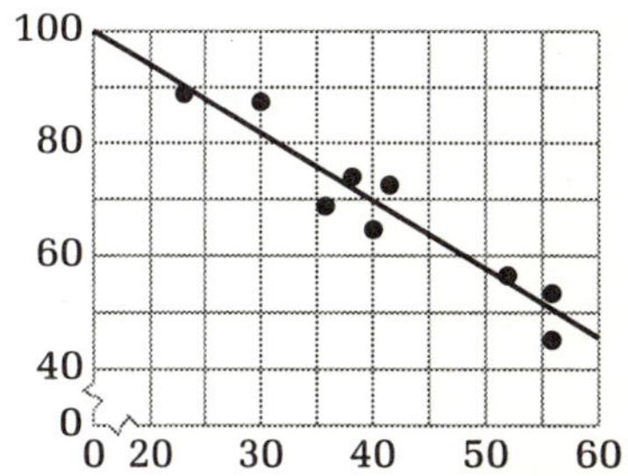

UNDERSTANDING THE MAIN IDEAS

Using a scatter plot of data to make predictions

Sample 1

This table shows the average height and weight for boys from the ages of 1 to 10 years of age.

Age	1	2	3	4	5	6	7	8	9	10
Height (in.)	17	23	36	39	42	45	47	50	52	54
Weight (lb)	21	26	31	34	39	46	51	57	63	69

What average weight could be predicted for a boy that is 60 inches tall?

Sample Response

Draw a scatter plot of the data.

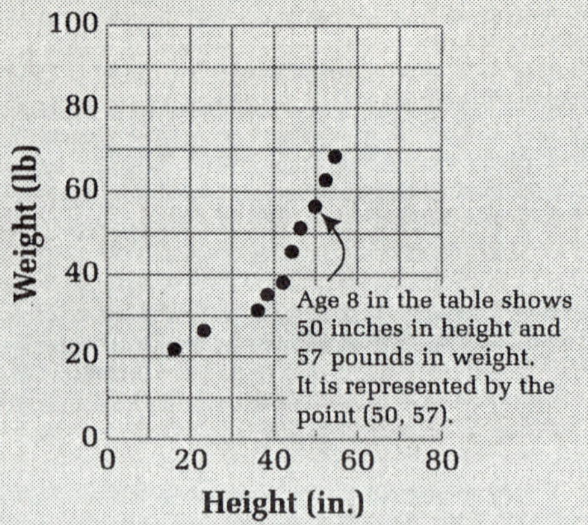

Now draw a fitted line. (The line must lie close to the points on the scatter plot, but there is not one correct placement.) Then use the fitted line to make a prediction.

A boy 60 inches tall would weigh about 77 pounds.

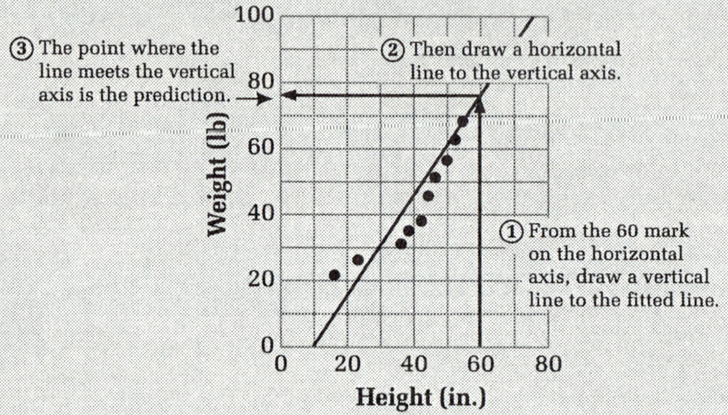

1. The table below shows the heights and weights of girls from 1 to 10 years of age. What average weight could be predicted for a girl 60 in. tall?

Age	1	2	3	4	5	6	7	8	9	10
Height (in.)	17	21	36	39	41	44	47	50	52	54
Weight (lb)	20	25	30	33	38	45	49	56	62	69

Sample 2

State whether each scatter plot shows a *positive correlation*, a *negative correlation*, or *no correlation*.

a. 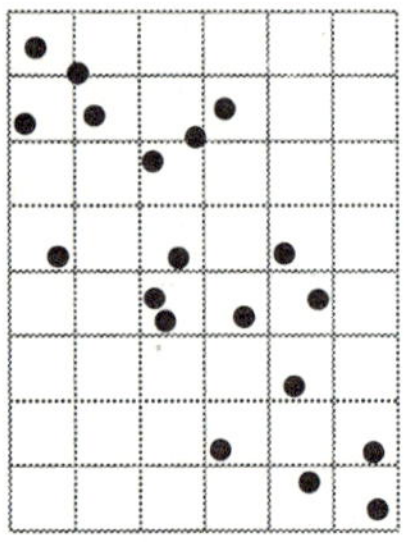

b. 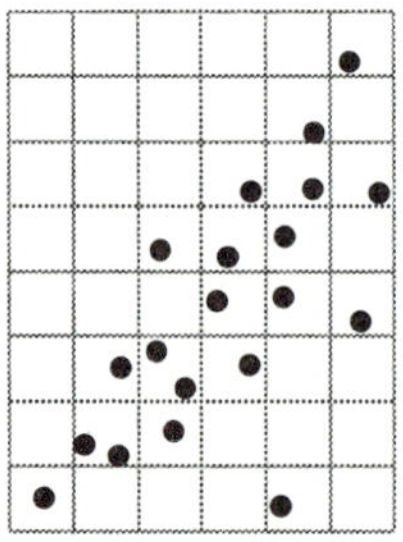

c. 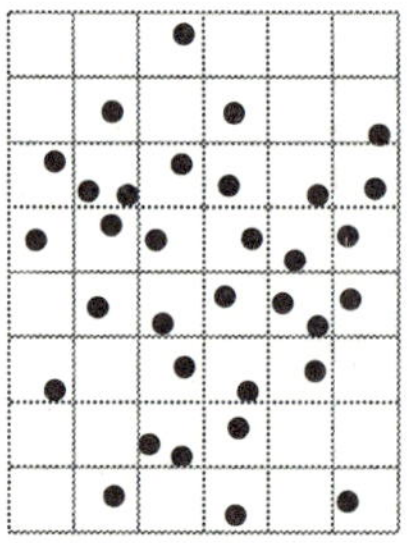

Sample Response

a. Notice that as the points move left to right, they also move from the top to the bottom. Therefore, this scatter plot shows a *negative correlation*.

b. As these points move from left to right, the points also move from bottom to top. This scatter plot shows a *positive correlation*.

c. As the points on this scatter plot move from left to right there is no indication of the points moving up or down. This scatter plot shows *no correlation*.

State whether the following scatter plots show a *positive correlation*, a *negative correlation*, or *no correlation*.

2.

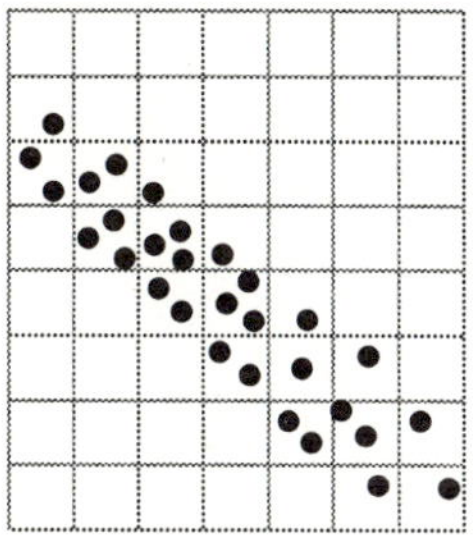

3.

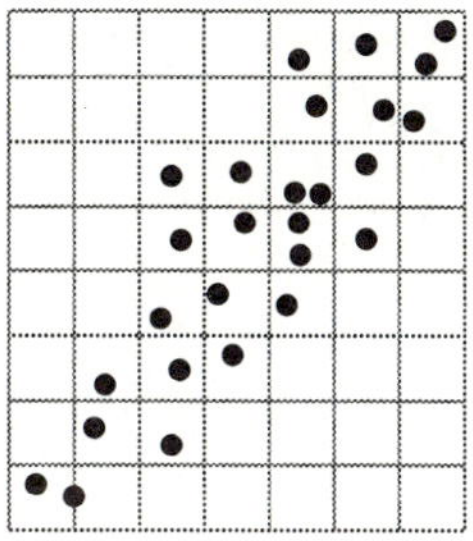

Sample 3

Make a scatter plot for the elevation and the total 1990 precipitation. State whether the scatter plot shows a *positive correlation*, a *negative correlation*, or *no correlation*.

City	Latitude (°N)	Elevation (ft)	Total 1990 Precipitation (in.)	Total 1990 sleet/snow (in.)
Anchorage, AK	61	114	19.0	95.9
Buffalo, NY	42	705	50.9	67.2
Seattle, WA	47	400	44.8	13.6
Omaha, NE	41	997	26.3	27.2
Grand Rapids, MI	43	784	42.6	60.0
Jackson, MS	32	291	54.0	0.0
Fresno, CA	36	328	8.7	0.0
Minneapolis, MN	45	834	33.1	33.9
St. Louis, MO	38	535	45.1	28.7
Tampa, FL	28	19	34.4	0.0

Sample Response

To make a scatter plot for the elevation and the total 1990 precipitation, ordered pairs must be plotted. The ordered pairs for this sample will be in the form (elevation, total precipitation).

Elevation (ft)	Total 1990 Precipitation (in.)	Ordered Pair
114	19.0	(114, 19.0)
705	50.9	(705, 50.9)
400	44.8	(400, 44.8)
997	26.3	(997, 26.3)
784	42.6	(784, 42.6)
291	54.0	(291, 54.0)
328	8.7	(328, 8.7)
834	33.1	(834, 33.1)
535	45.1	(535, 45.1)
19	34.4	(19, 34.4)

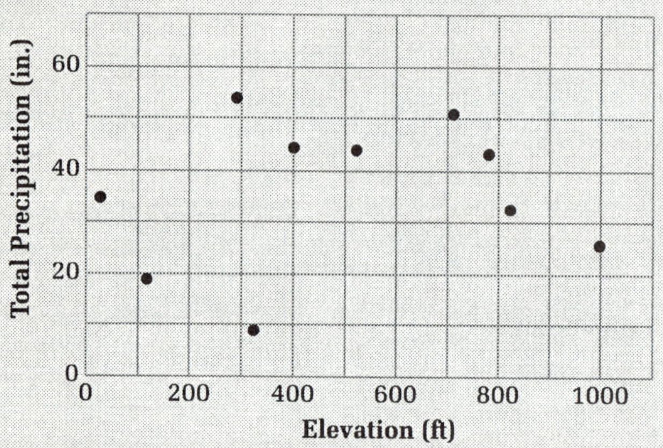

The random arrangement of dots in this scatter plot indicate that there is *no correlation* between elevation and total precipitation.

Make a scatter plot for each relationship using the information given in the table in Sample 3. State whether each scatter plot shows a *positive correlation*, a *negative correlation*, or *no correlation*.

4. the latitude and the total precipitation

5. the latitude and the total snow/sleet

Review **PREVIEW**

For Exercises 6–8, graph each inequality on a number line.

6. $a \geq 7$

7. $x \leq -3$

8. $-4 \leq k \leq 5$

9. Scott earns $4.25/h as a custodian in a school. Tell how much he earns for each number of hours he works.

a. 6 hours

b. 4 hours

c. y hours

Histograms and Stem-and-Leaf Plots

FOCUS

Use histograms, frequency tables, and stem-and-leaf plots to display data.

When the data from the frequency table below is entered in a computer graphics program, the graph shown below results.

Vehicle Speed (mph)	Number of Cars
50–51	1
52–53	3
54–55	5
56–57	2
58–59	2
60–61	4
62–63	7
64–65	16
66–67	2
68–69	2

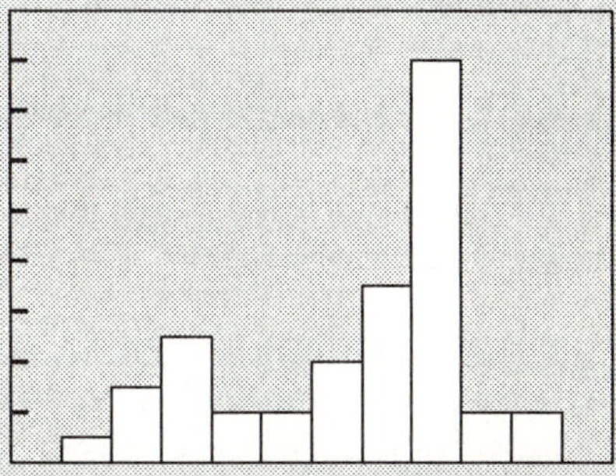

KEY TERMS | ***EXAMPLE / ILLUSTRATION***

Histogram
a type of bar graph that shows how many data items occur in each of one or more non-overlapping intervals

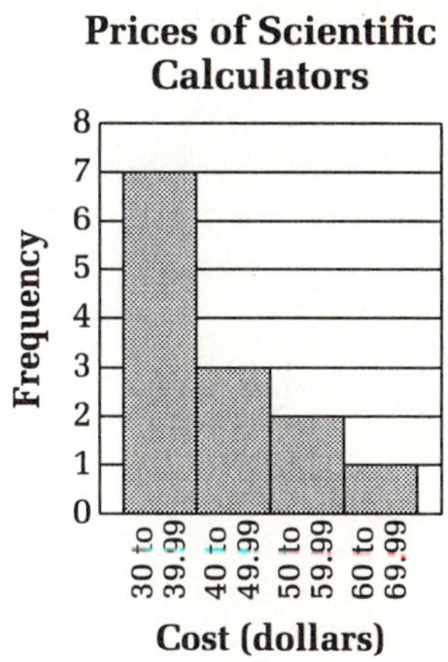

Frequency
the number of data items in an interval

In the histogram above, the frequency of scientific calculators in the $40 to $49.99 price range is 3.

Frequency table
a table that displays the exact number of data items in an interval

Prices of Scientific Calculators

Prices	Frequency
$30–39.99	7
$40–49.99	3
$50–59.99	2
$60–69.99	1

Stem-and-leaf plot a method of displaying data items by separating each item into two parts, a stem and a leaf	**Prices of Scientific Calculators** 3 | 0 0 2 3 5 9 4 | 0 0 7 5 | 0 0 6 | 4

UNDERSTANDING THE MAIN IDEAS

Frequency tables and histograms

A histogram is a useful way of displaying data that can be grouped into convenient intervals. The data used to make a histogram is often obtained from a frequency table. In a frequency table, the intervals are defined and the number of pieces of data in each interval is given.

Sample 1

The weights in pounds of 25 players on a high school junior varsity football team are listed below.

142 146 148 150 152 155 158 159 159 160 163 165 166
167 167 168 169 170 173 174 175 177 180 182 191

a. Make a frequency table. Use the intervals 140–149, 150–159, 160–169, 170–179, 180–189, and 190–199.

b. Use your frequency table from part (a) to make a histogram.

Sample Response

a.

Weights of Football Players (in pounds)	
Weight	**Frequency**
140–149	3
150–159	6
160–169	8
170–179	5
180–189	2
190–199	1

b.

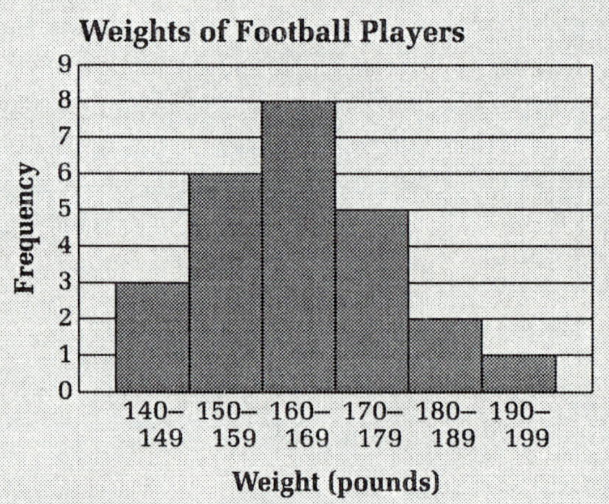

Thirty people were asked to tell how many books they had read in the past year. Here are the results of the survey:

0 2 4 6 6 8 10 12 12 13 14 14 14 14 15 15
15 16 16 18 19 19 20 21 22 24 30 36 48 48

1. Make a frequency table. Use the intervals 0–9, 10–19, 20–29, 30–39, and 40–49.
2. Use your frequency table from Exercise 1 to make a histogram.

Stem-and-leaf plots

A stem-and-leaf plot is a useful way of organizing data into intervals in such a way that each data item is displayed. On a stem-and-leaf plot, each data item is separated into a stem and a leaf. The leaf is the last digit of a data value and the stem is the remaining digit or digits. For example, the stem for the number 32 would be 3 and the leaf would be 2, whereas for the number 185 the stem would be 18 and the leaf would be 5.

Sample 2

Use the stem-and-leaf plot shown at the right.

a. What are the intervals represented by the stems of the data?

b. What is the frequency for each interval?

c. List the individual data items.

d. What information does the stem-and-leaf plot have that the frequency table for Sample 1 does not have?

Stem	Leaf
14	2 6 8
15	0 2 5 8 9 9
16	0 3 5 6 7 7 8 9
17	0 3 4 5 7
18	0 2
19	1

Sample Response

a. The stem gives the first two digits of each interval: 140–149, 150–159, 160–169, 170–179, 180–189, 190–199.

b. The *number* of leaves for a stem is the frequency of each interval. From top to bottom, the frequencies are 3, 6, 8, 5, 2, 1.

c. Combine each leaf with its stem to get the individual data items. The data items in the interval 140–149 are 142, 146, and 148; in the interval 150–159, the data items are 150, 152, 155, 158, 159, and 159; in the interval 160–169, the data items are 160, 163, 165, 166, 167, 167, 168, and 169; in the interval 170–179, the data items are 170, 173, 174, 175, and 177; for the interval 180–189, the data items are 180 and 182; and the only item in the interval 190–199 is 191.

d. The stem-and-leaf plot gives the individual data items, not just the intervals and the frequencies.

For Exercises 3 and 4, use the following test scores.

58	**60**	**63**	**65**	**65**	**67**	**69**	**70**	**70**	**71**
74	**75**	**75**	**75**	**78**	**81**	**81**	**83**	**84**	**84**
87	**88**	**90**	**91**	**91**	**93**	**93**	**95**	**96**	**98**

3. Make a stem-and-leaf plot of the test scores.

4. Make a histogram from your stem-and-leaf plot.

Write and graph an inequality to describe the interval referred to in each statement.

5. In Iowa, a person must be 16 years old in order to obtain a driver's license.

6. Water is in a liquid state between the temperatures of 0°C and 100°C.

Draw an angle whose measure is the given number of degrees.

7. 30° **8.** 120° **9.** 170°

10. Find the median, the range, and the outliers of the salaries listed below.

\$5400	\$18,500	\$32,000	\$12,400	\$29,500
\$16,200	\$25,000	\$18,500	\$30,000	\$27,300
\$20,400	\$28,800	\$55,000	\$33,000	\$18,500

Mean, Median, and Mode

FOCUS

Find values that are typical of a set of data, identify data values that are not typical of a data set, and use equations to find missing data values.

For certain kinds of data, such as heights or shoe sizes, if you collect data for a great many people and then graph the data, you will get a curve that looks something like the one shown below.

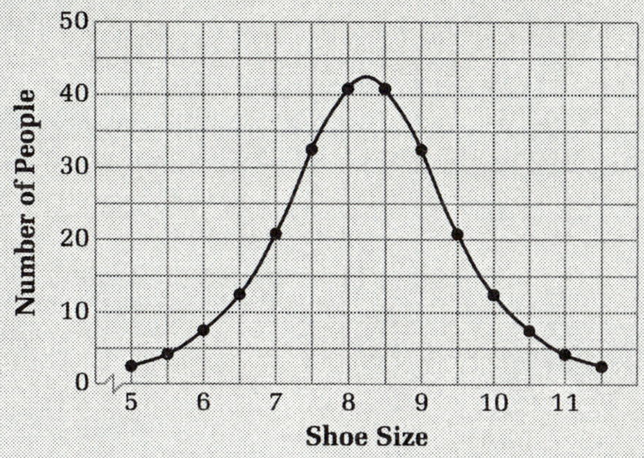

This bell-shaped curve is called a normal curve *and the distribution of data is called a* normal distribution. *In a normal distribution, the mean, median, and mode are all the same number.*

KEY TERMS	EXAMPLE / ILLUSTRATION
Mean the sum of the items in a set of data divided by the number of items	Data values: 32, 64, 64, 72, 73, 76, 80, 83, 89, 89, 92 The mean of the data is 74.
Median the middle number in a set of data after the data are put in order (If the number of data items is even, the median is the mean of the two middle numbers.)	The median of the data above (the data is in order) is 76.
Mode the number (or numbers) that occurs most often in a set of data (A data set may have more than one mode.)	There are two modes of the data above, 64 and 89.
Outlier data values that are much larger or smaller than most of the values in a data set (Outliers are *not* typical of the data set.)	There is one outlier in the data above, 32.
Range the difference between the smallest and largest numbers	The range of data above is 60.

UNDERSTANDING THE MAIN IDEAS

Mean, median, and mode

One value from a set of data is often chosen as the most typical value for describing the set. This value may be the mean, which you find by adding all the values and then dividing by the number of values in the set.

However, very large or very small values, called outliers, can make the mean a misleading value for describing the data. The typical value might be the median value, which is the middle value when the values are arranged in order. (If there are two "middle" numbers, the median is the mean of these two numbers.) Or the typical value might be the mode, the value (or values) that occurs most often.

Sample

The weekly salaries of the ten sales clerks at Howard's Book Store are \$375, \$350, \$440, \$400, \$400, \$430, \$650, \$350, \$410, and \$210.

a. Find the mean, median, and mode(s) of the salaries.

b. What is the salary range?

c. Are there any outliers?

Sample Response

a. mean: (375 + 350 + 440 + 400 + 400 + 430 + 650 + 350 + 410 + 210) ÷ 10 = 4015 ÷ 10 = 401.5; \$401.50

median: The ordered list of data is \$210, \$350, \$350, \$375, \$400, \$400, \$410, \$430, \$440, and \$650. The two middle numbers are both \$400, so the median is \$400.

mode: There are two modes, \$350 and \$400.

b. salary range: \$650 − \$210 = \$440

c. outliers: \$210 and \$650

For Exercises 1–6, use the following test scores.

45 62 66 66 68 71 71 71 72 74 75 75 76
77 80 80 80 81 83 84 84 86 89 90 95

1. What is the range of the scores?

2. What is the mean of the scores?

3. What is the median score?

4. What are the modes of the scores?

5. Are there any outliers? If so, what are they?

6. ***Writing*** Suppose the grades A, B, C, D, and F are to be assigned to the test scores. How would you group the scores to assign letter grades?

Review PREVIEW

7. You want to show what percent of the people surveyed chose one of five different vegetables as their favorite vegetable. Would you use a *circle graph* or a *line graph* to display the survey results? Explain.

Write each product as a power.

8. $6 \cdot 6 \cdot 6 \cdot 6 \cdot 6$

9. $y \cdot y \cdot y \cdot y$

Graph each pair of numbers on a number line. Then use the symbol < to write an inequality comparing the two numbers.

10. −4 and −3

11. 0 and −5

Box-and-Whisker Plots

FOCUS

Draw a box-and-whisker plot and use it to compare data sets.

Computer graphics are affecting the way in which data is displayed. Many examples of computer graphics can be found in newspapers and magazines. The diagrams below show the stages in changing a two-dimensional circle graph to a three-dimensional pie chart that emphasizes one part of the data.

Recycling Data | **Recycling Data** | **Recycling Data**

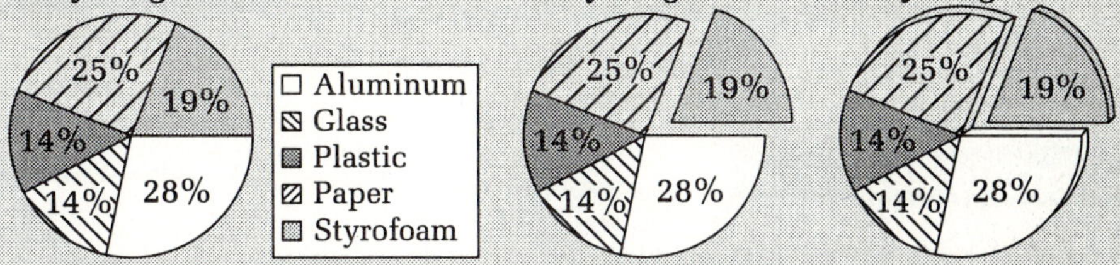

After you apply the format file, your circle graph looks like this.

Drag a slice of the graph to make it stand out from the rest of the graph.

This is the finished pie graph with depth added.

KEY TERMS	*EXAMPLE / ILLUSTRATION*
Box-and-whisker plot a graph that shows the median, the quartiles, the extremes, and the range of a set of data	The box-and-whisker plot for the data 43, 49, 51, 52, 55, 56, 59, 61, 64, 67, and 70 is shown below. 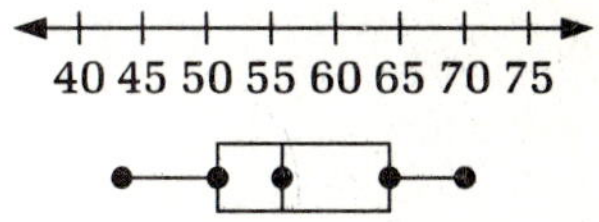
Lower quartile the median of the data in the lower half of a data set	The lower quartile of the data above is 51. It is represented as the left end of the box in the plot.
Upper quartile the median of the data in the upper half of a data set	The upper quartile of the data above is 64. It is represented as the right end of the box in the plot.
Extremes the lowest and highest data values in a data set (The lowest data value is the *lower extreme* and the highest data value is the *upper extreme.*)	The lower and upper extremes of the data above are 43 and 70, respectively. These extremes are shown as the endpoints of the whiskers in the plot.

Sample I

Eighteen basketball players practiced free throw shots on two separate days. The number of shots made out of 50 free throws by each player are shown below.

Day 1

12	32	42	50	39	38
27	34	42	35	48	36
13	15	23	16	25	43

Day 2

42	17	18	42	28	37
19	14	14	41	21	29
21	17	19	26	31	34

a. Make box-and-whisker plots for the two sets of data.

b. Compare the results for the two days of practice.

Sample Response

a. ***Step 1*** Write the data items in order for each data set.

Day 1: 12 13 15 16 23 25 27 32 34 35 36 38 39 42 42 43 48 50

Day 2: 14 14 17 17 18 19 19 21 21 26 28 29 31 34 37 41 42 42

Step 2 Find the median, the extremes, and the quartiles of each data set.

Day 1: median: There are two middle numbers, 34 and 35, so the median is (34 + 35) ÷ 2 = 34.5; extremes: 12 and 50; lower quartile: the middle number of the lower half of the data is 23; upper quartile: the middle number of the upper half of the data is 42.

Day 2: median: There are two middle numbers, 21 and 26, so the median is (21 + 26) ÷ 2 = 23.5; extremes: 14 and 42; lower quartile: the middle number of the lower half of the data is 18; upper quartile: the middle number of the upper half of the data is 34.

Step 3 Graph the five values for each data set found in Step 2 below a number line.

Step 4 Draw a box from the lower quartile to the upper quartile. Draw a vertical line in the box through the median.

Step 5 Draw whiskers (line segments) from the ends of the box to the extremes.

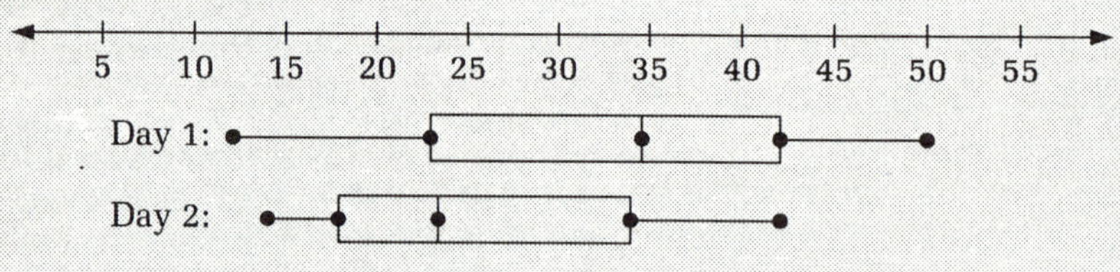

b. The Day 2 results show a much lower performance than the Day 1 results. The upper quartile for the second day was less than the median for the first day, and the median for the second day was only slightly better than the lower quartile for the first day. While the lower extreme was a little higher for the second day, the upper extreme for the second day was much less than that for the first day.

The daily high temperatures for each day of two one-week periods in Chicago in July are given below. Use these temperatures for Exercises 1 and 2.

Week 1:	**95**	**96**	**93**	**98**	**98**	**101**	**99**
Week 2:	**95**	**85**	**83**	**78**	**79**	**82**	**80**

1. Make box-and-whisker plots for the two sets of data.

2. Compare the weather in Chicago during each of these two weeks.

3. ***Open-ended*** Collect two sets of data about a situation. Make box-and-whisker plots and compare the data sets.

Review PREVIEW

For Exercises 4–6, use the histogram showing the prices of notebook computers.

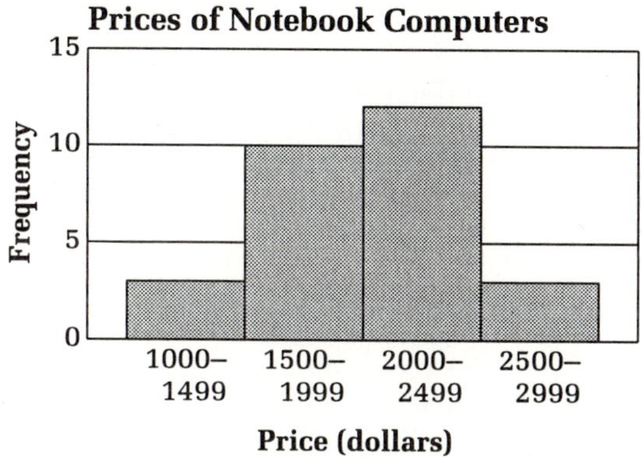

4. How many computer models cost \$2000 or more?

5. Which price range has the greatest frequency? What does this mean in terms of the data?

6. What else do you need to know to make a stem-and-leaf plot for the data set?

For Exercises 7 and 8, simplify.

7. $6x - 2y - 5x + 3y$

8. $-8a^2 + 6a^2$

9. A birdwatcher records the number of migrating warblers seen in a park each day over a two-week period. Is this data discrete or continuous? Should the birdwatcher use a bar graph or a line graph to display the data?

Investigating Probability

FOCUS

Describe how likely it is that something will happen.

Manufacturers must test their products to determine if they meet quality standards. For example, a sneaker manufacturer might take a sample of 100 pairs and inspect them closely to see if they meet all the company's standards. The manufacturer then uses the results of this inspection to estimate the probability that any pair of sneakers meets the standards.

KEY TERMS	EXAMPLE / ILLUSTRATION
Outcome a possible result of an experiment	A three-way race for Senator involving a Republican, a Democrat, and an Independent is an experiment with three possible outcomes: the Republican wins, the Democrat wins, or the Independent wins.
Event a particular outcome or combination of outcomes of an experiment	In the election described above, the event "the Independent loses" can happen in one of two ways (outcomes): the Republican wins, or the Democrat wins.
Probability a number from 0 to 1 (where 0 indicates impossibility and 1 indicates certainty) that describes how likely it is that an event will occur	Weather forecasts are given in probabilities (or chances). A "70% chance of rain" means that the probability of rain is 0.70.
Experimental probability probability that is based on the results of an experiment, an activity in which data is observed and recorded	Researchers find that a medicine they are testing improved the condition of 270 out of 300 patients. Using experimental probability, they conclude that the probability the medicine will help a patient who takes it is $\frac{270}{300} = 0.9$, or 90%.
Theoretical probability probability found without having to do an experiment, based on the assumption that all outcomes of an experiment are equally likely to occur	You buy 2 tickets for a charitable lottery. Eighty-five tickets in all are sold. Since each ticket is equally likely to be drawn, you can use theoretical probability to reason that your chances of winning are $\frac{2}{85}$, or about 0.024.

Complementary events two events, the event E and the event "E does not happen" (The sum of the probabilities of complementary events is 1.)	In a lottery, the events "you win" and "you lose" are complementary.

UNDERSTANDING THE MAIN IDEAS

Experimental probability

The experimental probability of an event E can be calculated using the formula

$$P(E) = \frac{\text{number of times event } E \text{ occurs}}{\text{number of times the experiment is done}}.$$

The more times the experiment is conducted, the more reliable is the computed value of $P(E)$.

Sample 1

A vending machine dispenses five different types of fruit drinks. The owner of the machine has kept records on the last 2000 sales from the machine.

Fruit punch	746
Citrus punch	524
Apple juice	98
Orange juice	282
Cranberry juice	350

What is the probability that a randomly-chosen purchase from the machine will be a punch?

Sample Response

Of the five types of juice, two are punches: fruit punch and citrus punch.

$$P(\text{purchasing a punch}) = \frac{\text{number of times a punch was purchased}}{\text{number of times a purchase was made}}$$

$$= \frac{746 + 524}{2000} \longleftarrow$$ Add the number of fruit punches and citrus punches.

$$= \frac{1270}{2000}$$

$$= 0.635$$

The probability that a punch is bought is 0.635, or $63\frac{1}{2}\%$.

For Exercises 1 and 2, use the information provided in Sample 1.

1. Find the probability that a randomly-selected customer buys
   a. apple juice.
   b. apple juice or cranberry juice.
   c. seltzer water.

2. Of the next 50 drinks sold, how many are likely to be orange juice?

Theoretical probability

When all the possible outcomes of an experiment are equally likely, the following ratio is used to determine the theoretical probability of an event E.

$$P(E) = \frac{\text{the number of outcomes that are in event } E}{\text{the number of possible outcomes}}$$

If you know the probability of an event E, then the probability that event E does *not* happen is $1 - P(E)$.

Sample 2

The letters of the alphabet are printed on cards (one letter per card) and placed in a box. You close your eyes and choose one card. What is the probability that the card you chose is

a. the first letter of your first name?

b. a vowel?

c. *not* a vowel?

Sample Response

Each card is equally likely to be chosen, so theoretical probability can be used.

a. Of the 26 cards, only one has your initial on it.

$$P(\text{choosing your first initial}) = \frac{1}{26}$$

b. There are five vowels (A, E, I, O, U), so the event "choosing a vowel" can happen in 5 ways.

$$P(\text{choosing a vowel}) = \frac{5}{26}$$

c. "Choosing a vowel" and "not choosing a vowel" are complementary events.

$$\begin{aligned} P(\text{not choosing a vowel}) &= 1 - P(\text{choosing a vowel}) \\ &= 1 - \frac{5}{26} \\ &= \frac{21}{26} \end{aligned}$$

3. In Sample 2, what is the probability that Maria Carvalho chooses one of her two initials? that Alan Adams chooses one of his initials?

4. What is the probability that the last digit of your teacher's phone number is "1?"

5. Which of these pairs of events are complementary?

a. passing Math, failing Math

b. getting a traffic ticket, driving safely

c. reading, watching television

For Exercises 6–9, use the spinner at the right. Find each probability.

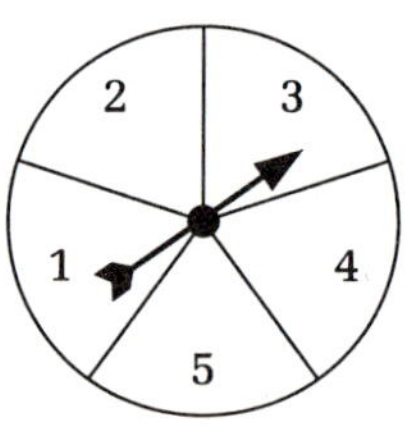

6. P(odd number) **7.** P(number less than 3)

8. P(3) **9.** P(not 4)

10. Using the spinner at the right, describe two complementary events.

Review PREVIEW

Write each ratio as a fraction in simplest form.

11. 10 : 55 **12.** 82 to 100 **13.** 18 : 48

Find the square root and cube root of each number. Round decimal answers to the nearest hundredth.

14. 36 **15.** 125 **16.** 91

Solve each equation.

17. $13x = 39$ **18.** $\frac{x}{1.7} = 4$ **19.** $\frac{x}{-6} = \frac{8}{3}$

Geometric Probability

FOCUS

Use lengths and areas to determine theoretical geometric probabilities.

Bales of hay are to be dropped from an airplane by parachute to cattle stranded by a blizzard. The cattle are clustered together within the inner circle of the drop area shown at the right. The probability that a bale of hay will land within the circular region where the cattle are located can be found by comparing the area of the small circle to the area of the large circle.

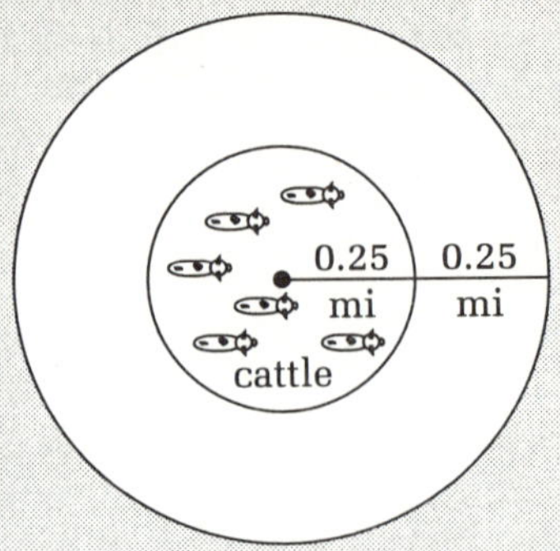

drop area

KEY TERMS

Geometric probability
probability based on areas and lengths

EXAMPLE / ILLUSTRATION

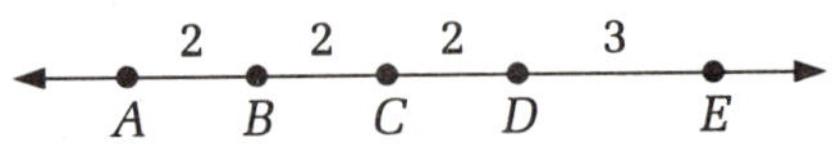

The probability that a point chosen at random on $\overline{AE}$ will be on $\overline{BC}$ is $\dfrac{\text{length of } \overline{BC}}{\text{length of } \overline{AE}} = \dfrac{2}{9}$.

UNDERSTANDING THE MAIN IDEAS

Theoretical geometric probability

You can use what you know about probability as the ratio of successful outcomes to possible outcomes to find probabilities related to areas and lengths of geometric figures.

Sample 1

Actors trying out for roles in a play are randomly assigned 20-min time slots between the hours of 1:00 P.M. and 4:00 P.M. What is the probability that an actor will be assigned the time slot from 2:20 P.M. to 2:40 P.M.?

Sample Response

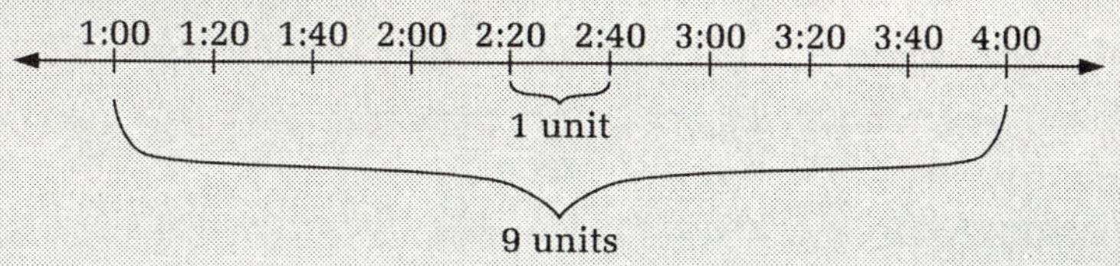

Find the ratio of the length of the segment for the 2:20 to 2:40 time slot to the length of the segment for the entire audition period.

$$\text{probability} = \frac{\text{length of segment for 2:20 - 2:40}}{\text{length of segment for 1:00 - 4:00}} = \frac{1}{9}$$

The probability of being assigned the 2:20 P.M. to 2:40 P.M. time slot is $\frac{1}{9}$.

Sample 2

In the situation about the stranded cattle described on the previous page, what is the probability that a bale of hay will land within the region where the cattle are clustered?

Sample Response

Make a sketch. Use the formula $A = \pi r^2$ to find the areas of the two circles. The probability is the ratio of the two areas.

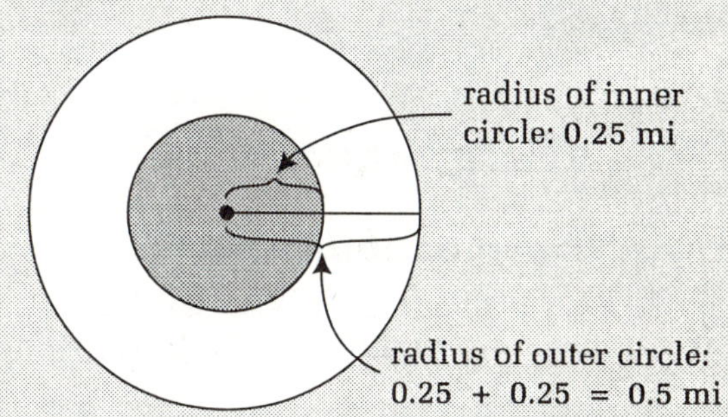

$$\text{probability} = \frac{\text{area of inner circle}}{\text{area of outer circle}}$$
$$= \frac{\pi(0.25)^2}{\pi(0.5)^2}$$
$$= \frac{0.0625}{0.25}$$
$$= \frac{1}{4}$$

For each bale of hay, the probability of landing within the region where the cattle are clustered is $\frac{1}{4}$.

1. Which would be greater, the probability that a point chosen at random from each square is in the shaded region in square **A** or in the shaded region in square **B**?

A.

B.

2. Mita does her homework between 4:30 P.M. and 6:00 P.M. each day. She usually spends 30 min completing her math assignment. If she receives a phone call during her homework time, what is the probability that she will be working on math?

3. What is the probability that a mosquito flying at random against this screen will enter through the triangular tear in the screen?

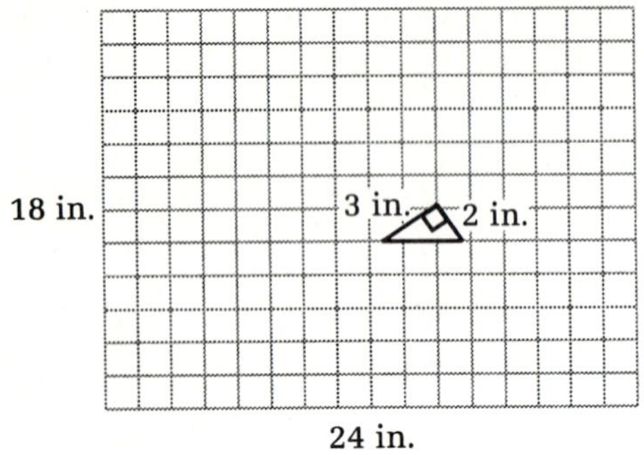

4. The numbers in a calculator display are formed by lighting up various segments of the array shown at the right. If you enter a digit at random, what is the probability that segment *d* will be lit? (*Hint:* Check your calculator to see which segments are lit for each digit.)

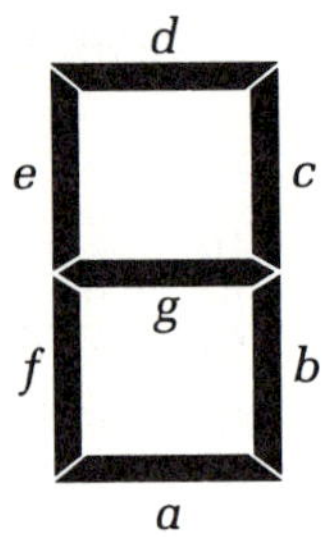

Review **PREVIEW**

5. **a.** Write the converse of the statement "If $\sqrt{12} = x$, then $12 = x^2$."
   b. Tell whether the converse is *True* or *False*. If it is false, find a counterexample.

6. Suppose a figure is a regular triangle. What are three things you know about the figure?

Give each formula.

7. area of a trapezoid

8. area of a circle

Graphs and Functions

FOCUS

Understand what a function is, identify control variables and dependent variables, draw graphs of functions, and recognize functions.

Small-market economics rely on supply and demand. This graph of a demand curve illustrates that the price of wheat depends on the quantity available. As the amount of wheat needed to satisfy the demand increases, the price of the wheat decreases.

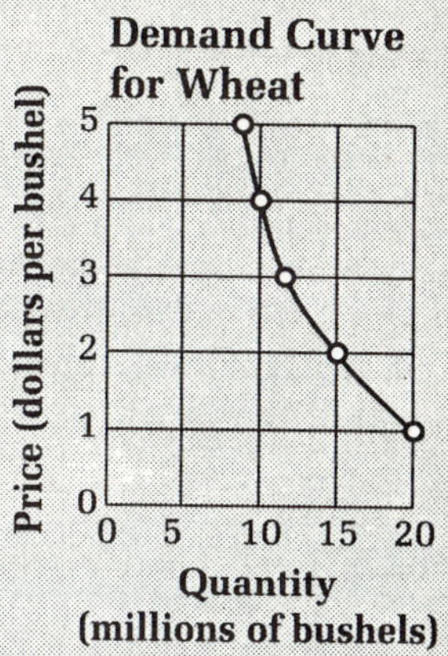

KEY TERMS | **EXAMPLE / ILLUSTRATION**

Dependent variable
a variable whose value is determined by one or more other variables

Control variable
a variable which determines the value of another variable (The control variable is sometimes called the *independent variable.*)

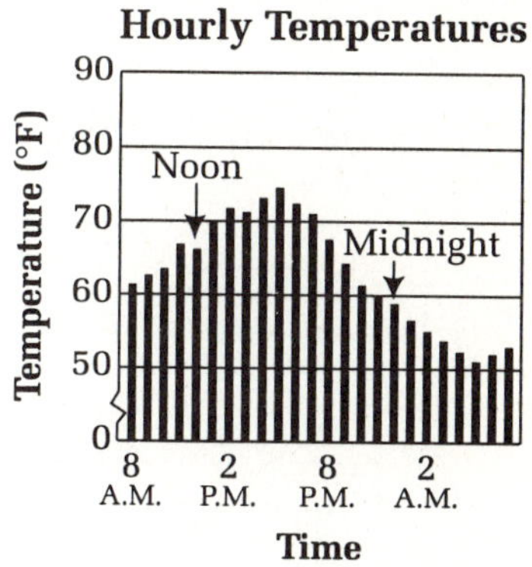

By examining this bar graph of the hourly temperatures starting from 8 A.M. one day to 7 A.M. the following day, it can be seen that the temperature depends on the time of day. The temperature is the *dependent variable* and the time of day is the *control variable.* Notice also that the time of day is on the horizontal axis and the temperature is on the vertical axis. In general, the control variable and the dependent variable are placed on those respective axes.

Function
a relationship in which there is only one value of the dependent variable for each value of the control variable

In the graph above, each hour has only one temperature. This is an example of a function.

UNDERSTANDING THE MAIN IDEAS

Functions

A function can be represented many ways: in words, in symbols, with a table of values, or with a graph.

Sample 1

At a week-long "going out of business" sale, the price of each item is one third the previous day's price. (Of course, if a customer waits until tomorrow, what they want may be gone.) Draw a graph of the price of an $486 item during the sale. Tell whether the graph represents a function.

Sample Response

First make a table of values.

Day of Sale	Price (dollars)
1	486
2	$\frac{1}{3}(486) = 162$
3	$\frac{1}{3}(162) = 54$
4	$\frac{1}{3}(54) = 18$
5	$\frac{1}{3}(18) = 6$
6	$\frac{1}{3}(6) = 2$
7	$\frac{1}{3}(2) = 0.67$

← The original price is $486.

← Multiply the previous day's price by $\frac{1}{3}$.

Then draw a graph. In general, the dependent variable is put on the vertical axis and the control variable on the horizontal axis. The price depends on the day of the sale. So the price is the dependent variable and the day of the sale is the control variable.

Daily Sale Price of an Item Originally Priced at $486

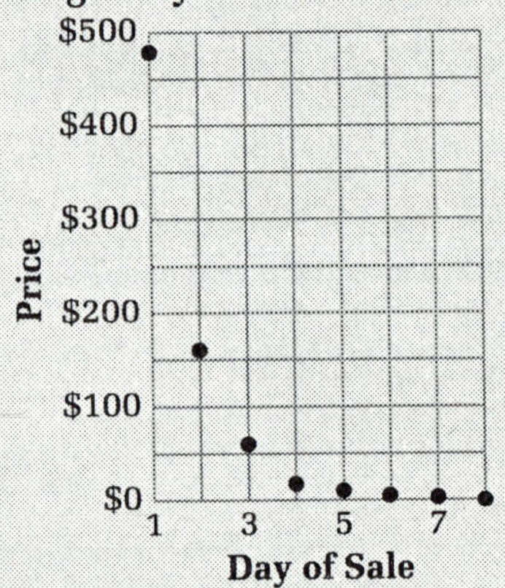

For each day of the sale, the item has only one price. So the graph represents a function.

Identify the dependent variable and the control variable in each situation.

1. Bob rides his bike at an average speed of 12 mph. When he rides for different lengths of time throughout the week, he travels different distances.

2. Karla walks or runs for 1 hour each day. She travels different distances each day according to her speed.

Draw a graph for each situation. Tell whether the graph represents a function. Explain.

3. The after-prom committee has a magazine sale every year. This year, they raised $500. They decide to try to raise 1.5 times the amount they raised the previous year for the next four years.

4. Alice earned these scores on her first five tests: 86, 75, 90, 75, and 95.

Vertical-line test for functions

A graph can be used to visualize a function. For each value of the control variable, there should be only one value of the dependent variable. This means that no two points on the graph of a function lie directly above one another. The vertical-line test involves imagining a vertical line drawn through a graph. If the vertical line passes through more than one point on the graph, then the graph does *not* represent a function.

Sample 2

Tell whether the graph represents a function. Explain your answer.

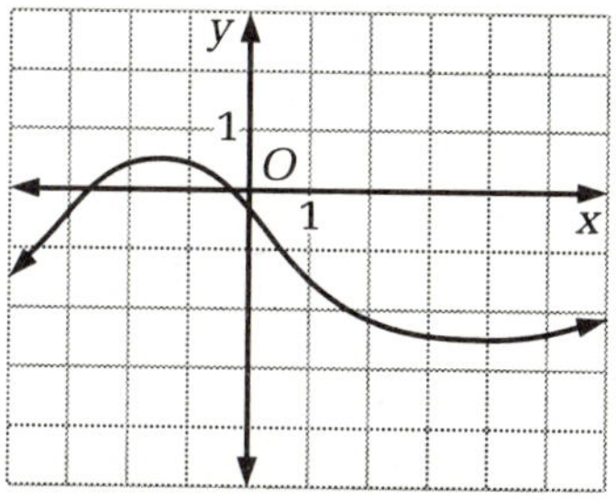

Sample Response

Imagine drawing numerous vertical lines through the graph.

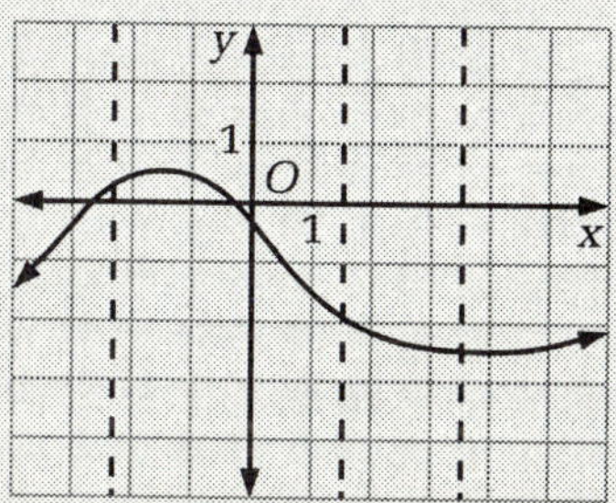

Since no two points of the graph lie on the same vertical line, the graph represents a function.

Tell whether each graph represents a function. Explain your answer.

5. 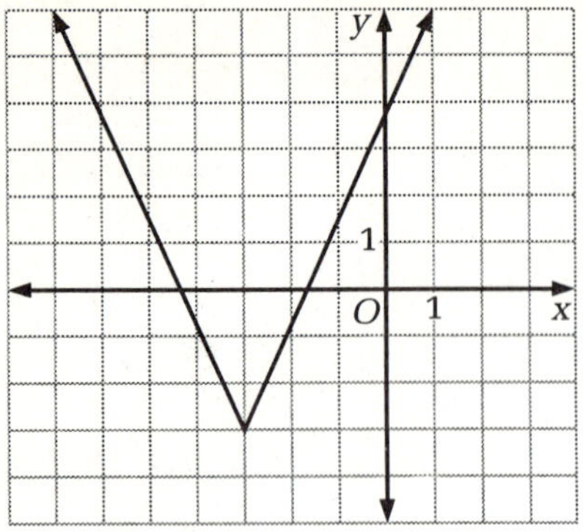

6. 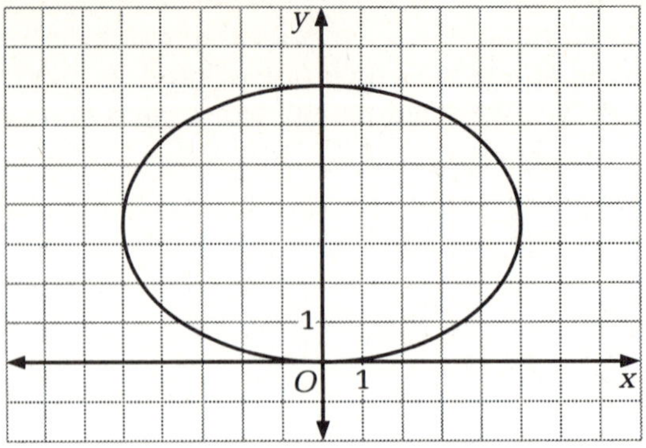

Sample 3

Describe the following situation using the phrase "is a function of." What is the dependent variable? the control variable?

Sam owns a house cleaning business. He charges different prices for cleaning houses of different sizes.

Sample Response

The cost depends on the size of the house. The cost is a function of the house size. The cost is the dependent variable. The house size is the control variable.

Describe each situation using the phrase "is a function of." What is the dependent variable? the control variable?

7. Mark plays golf every day. The more he plays, the lower his score.

8. Martha earns money for each craft item she assembles. The more she assembles, the more money she earns.

Review **PREVIEW**

9. Draw a scatter plot that shows a strong negative correlation.

Give the reciprocal of each number.

10. 6 **11.** $\frac{3}{4}$ **12.** $-\frac{5}{16}$

Evaluate each expression when $x = -5$, $x = 0$, and $x = 2$.

13. $-3x^2$ **14.** $|x| - 3$ **15.** $\frac{1}{x + 1}$

Exploring Angle Relationships

FOCUS

Investigate and use the relationships among angles in circles, angles in triangles, and angles formed by intersecting lines.

To be a successful billiards player, one must have the necessary playing skills, the ability to handle pressure, and an understanding of angles.

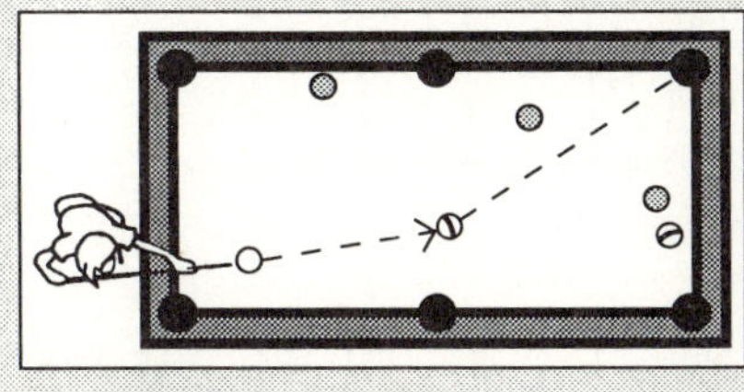

KEY TERMS	EXAMPLE / ILLUSTRATION
Ray a part of a line that extends forever in one direction	 ray *ST*, or $\overrightarrow{ST}$
Angle the figure formed when two rays meet at a common endpoint	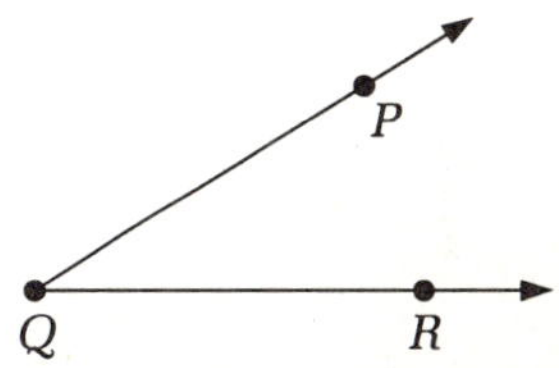 angle *PQR* or angle *RQP*, or $\angle PQR$ or $\angle RQP$
Vertex of an angle the common endpoint of the two rays that form an angle	The vertex of $\angle PQR$ above is the point *Q*.
Central angle of a circle an angle with its vertex at the center of a circle	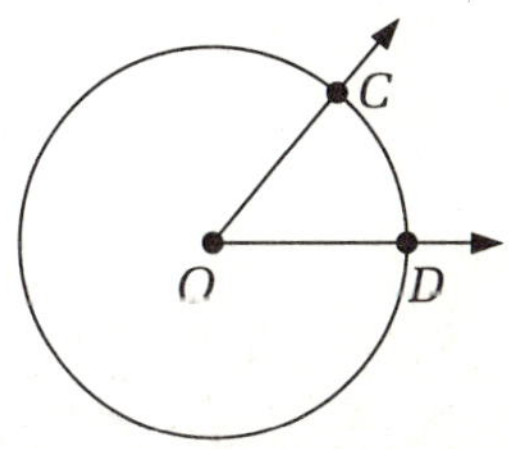 central angle *COD*
Right angle an angle whose measure is 90°	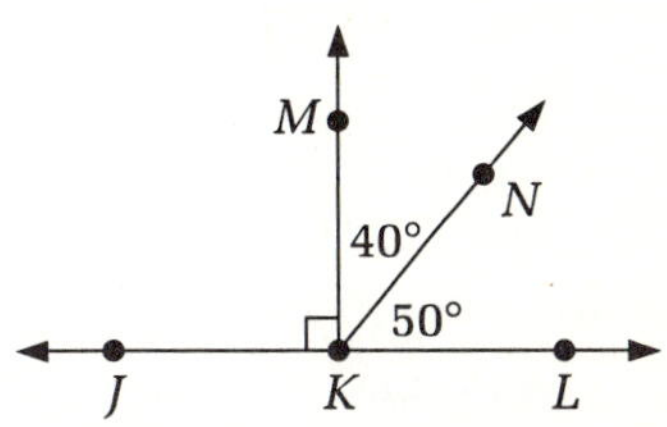 $\angle JKM$ and $\angle LKM$ are both right angles.
Straight angle an angle whose measure is 180°	In the figure above, $\angle JKL$ is a straight angle.

Obtuse angle
an angle whose measure is between 90° and 180°

In the figure for right angle, $\angle JKN$ is an obtuse angle.

Acute angle
an angle whose measure is between 0° and 90°

In the figure for right angle, $\angle MKN$ and $\angle NKL$ are both acute angles.

Complementary angles
two angles whose measures add up to 90°

In the figure for right angle, $\angle MKN$ and $\angle NKL$ are a pair of complementary angles.

Supplementary angles
two angles whose measures add up to 180°

In the figure for right angle, $\angle JKN$ and $\angle NKL$ are a pair of supplementary angles.

Congruent angles
angles with equal measures (The symbol ≅ means "is congruent to.")

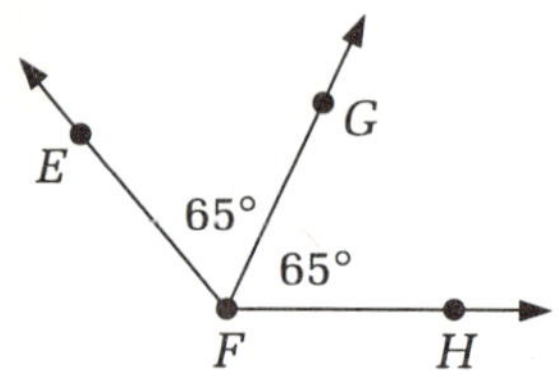

$\angle EFG \cong \angle GFH$

Vertical angles
two angles formed by intersecting lines and facing in opposite directions (Vertical angles have equal measures.)

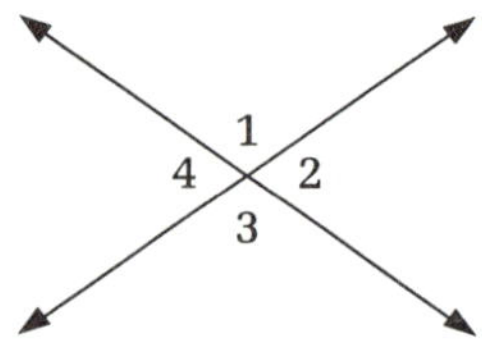

$\angle 1$ and $\angle 3$ are vertical angles.
$\angle 2$ and $\angle 4$ are vertical angles.

Acute triangle
a triangle with three acute angles

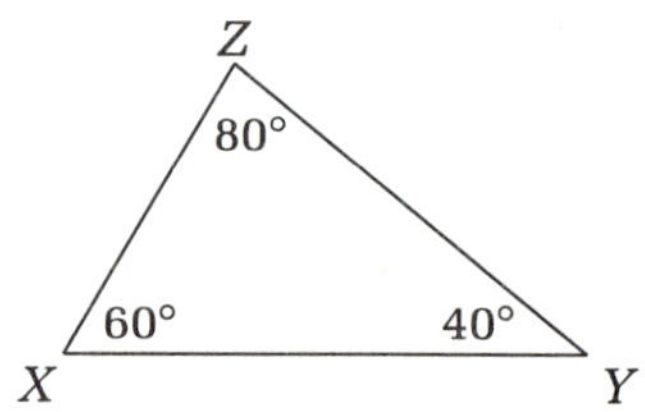

Obtuse triangle
a triangle with one obtuse angle

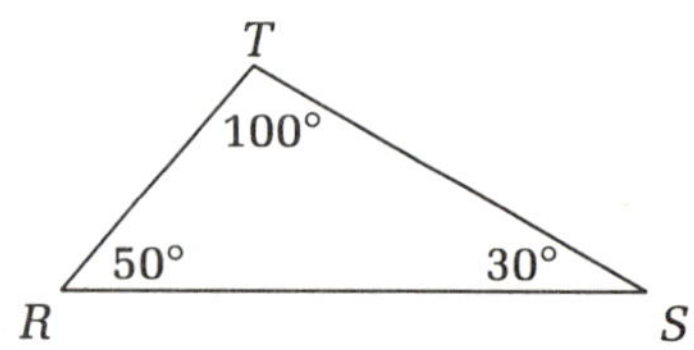

Right triangle
a triangle with one right angle

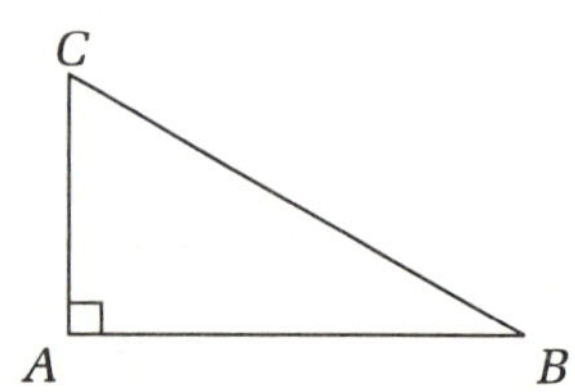

UNDERSTANDING THE MAIN IDEAS

Naming and classifying angles

An angle is a geometric figure that is formed when two rays meet at a common endpoint. The common endpoint is the vertex of the angle. An angle is usually named by using a point on each ray and the vertex, with the letter for the vertex between the letters for the other two points. Angles are measured in degrees and are classified by their measures. An angle with its vertex at the center of a circle is a central angle of the circle. Pairs of angles are complementary if the sum of their measures is 90°. The sum of the measures of two supplementary angles is 180°.

Sample 1

Use the figure at the right.

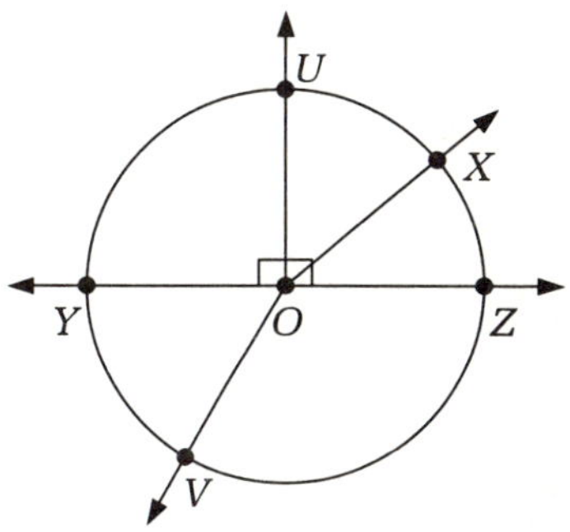

a. What point is the vertex of the angles shown?
b. What rays form $\angle VOY$?
c. Name three central angles of the circle.
d. Name two right angles.
e. Name three acute angles.
f. Name two obtuse angles.
g. Which angle is a straight angle?
h. Name a pair of complementary angles.
i. Name a pair of supplementary angles.

Sample Response

a. point O
b. $\overrightarrow{OV}$ and $\overrightarrow{OY}$
c. Answers may vary. Sample answers are given: $\angle VOY$, $\angle YOU$, and $\angle XOZ$.
d. $\angle YOU$ and $\angle UOZ$
e. $\angle XOZ$, $\angle VOY$, and $\angle UOX$
f. Answers may vary. Sample answers are given: $\angle YOX$ and $\angle XOV$.
g. $\angle YOZ$
h. $\angle XOZ$ and $\angle UOX$
i. Answers may vary. A sample answer is given: $\angle YOX$ and $\angle XOZ$.

Use the figure at the right for Exercises 1–6.

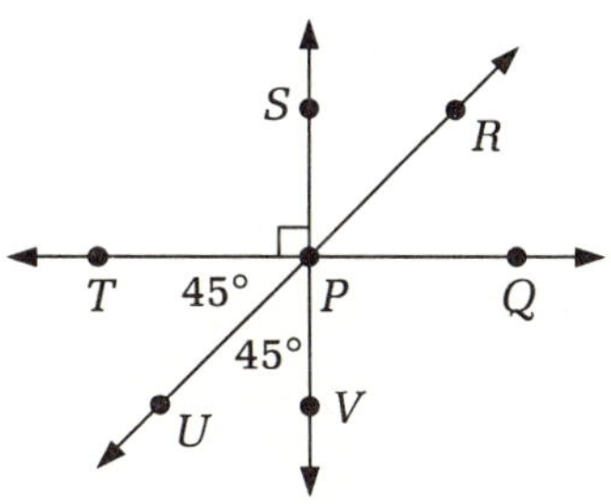

1. What rays form $\angle TPU$?
2. Name two acute angles.
3. Name three obtuse angles.
4. Name a pair of complementary angles.
5. Name a pair of supplementary angles.

Special angle relationships

Angles with equal measures are congruent. The pairs of congruent angles formed by intersecting lines are called vertical angles. The sum of the measures of the three angles of a triangle is always 180°. A triangle with three acute angles is an acute triangle, a triangle with one right angle is a right triangle, and a triangle with one obtuse angle is an obtuse triangle.

Sample 2

Use the figure at the right.

a. What is the measure of $\angle ACB$?

b. What is the measure of $\angle ACM$?

c. Name an obtuse triangle.

d. Name two pairs of vertical angles.

e. $\angle ACZ \cong$ ___?___

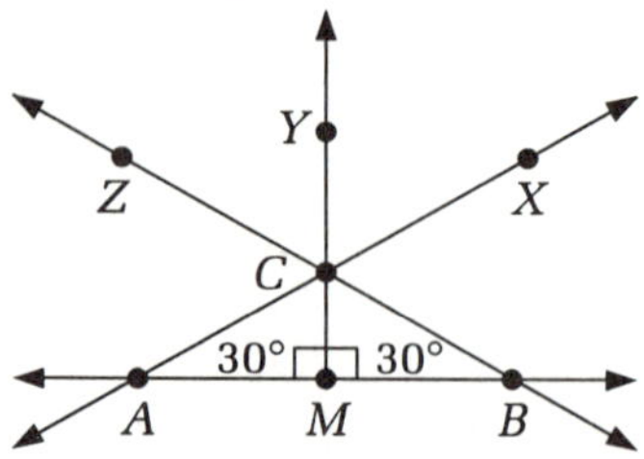

Sample Response

a. In $\triangle ABC$, $\angle ACB + 30 + 30 = 180$; $\angle ACB + 60 = 180$; $\angle ACB = 120°$

b. In $\triangle ACM$, $\angle ACM + 30 + 90 = 180$; $\angle ACM + 120 = 180$; $\angle ACM = 60°$

c. $\triangle ACB$

d. Answers may vary. Sample answers are given: $\angle ZCA$ and $\angle XCB$, $\angle ZCX$ and $\angle ACB$.

e. Since $\angle ACZ$ and $\angle XCB$ are vertical angles, $\angle ACZ \cong \angle XCB$.

For Exercises 6–9, write and solve an equation to find each unknown angle measure.

6.

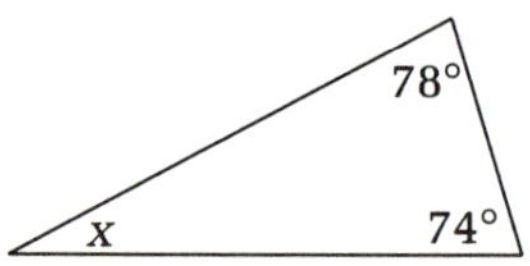

7.

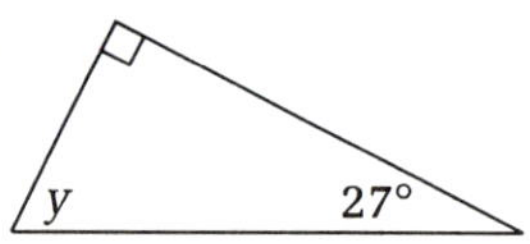

8.

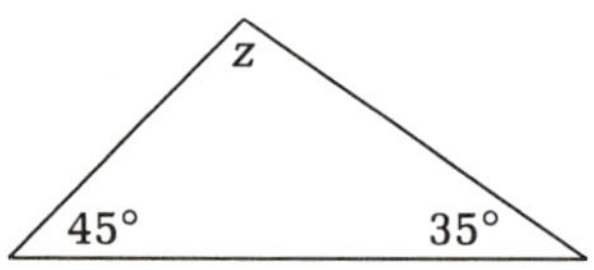

9.

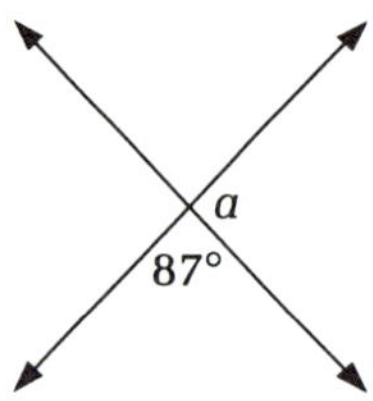

10. Classify each triangle in Exercises 6–8 as *right*, *acute*, or *obtuse*.

Review **PREVIEW**

Describe in words what each symbol means.

11. $\angle A$ **12.** $\cong$ **13.** $\overrightarrow{OP}$ **14.**

Name the property shown.

15. $(6 + 28) + 12 = 6 + (28 + 12)$

16. $32(12 + 8) = 32 \cdot 12 + 32 \cdot 8$

Simplify.

17. $(x - 5)7$

18. $5x^2 + 3 - 2x^2 - 8$

Working Together on Congruent Polygons

FOCUS

Create and identify congruent polygons.

The Mies van der Rohe "glass houses" on Lake Shore Drive in Chicago are twin apartment buildings whose matching faces are congruent rectangles. The rectangles outlined in the architectural floor plan below show the lobbies of the twin buildings.

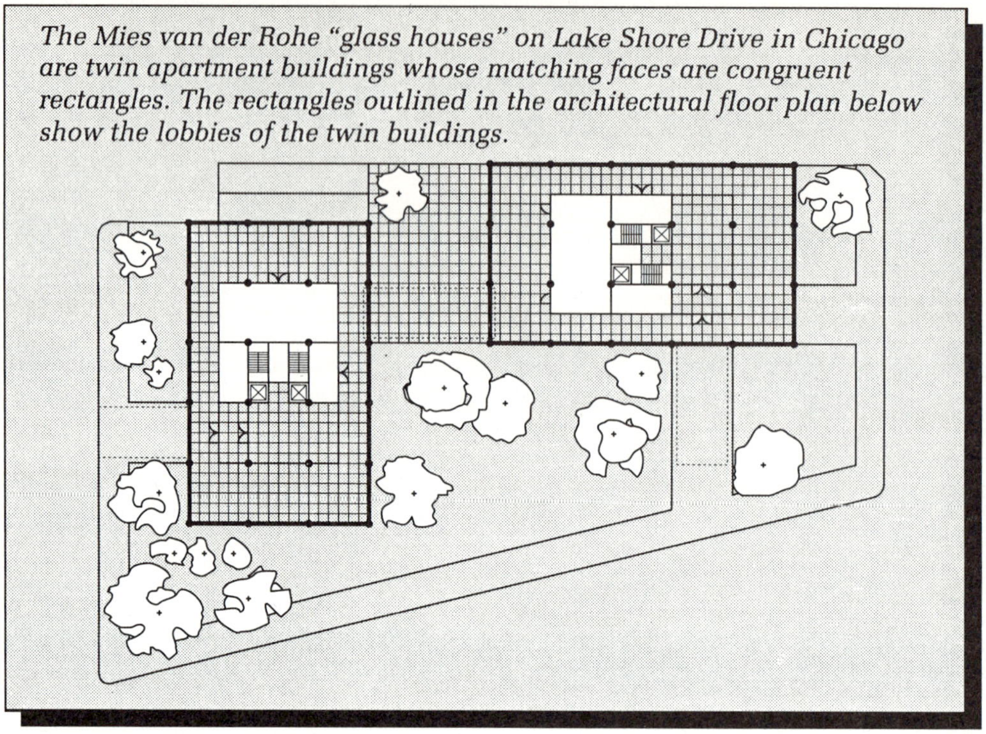

KEY TERMS	EXAMPLE / ILLUSTRATION
Congruent having the same size and shape (The symbol "≅" is read "is congruent to.")	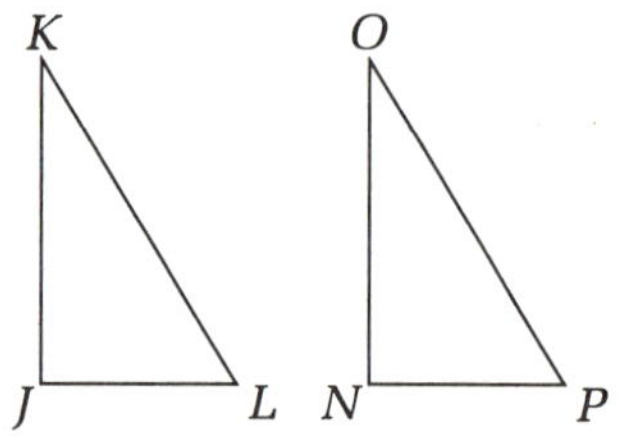 triangle *JKL* ≅ triangle *NOP*
Polygon a figure with straight sides, such as a triangle or a quadrilateral	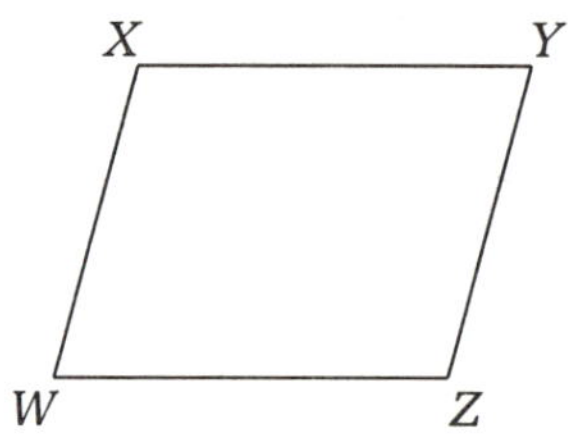 polygon *WXYZ*
Vertex a corner of a polygon (plural: *vertices*)	The vertices of quadrilateral *WXYZ* above are *W*, *X*, *Y*, and *Z*.
Side a part of a polygon defined by two consecutive vertices	Quadrilateral *WXYZ* above has sides *WX*, *XY*, *YZ*, and *WZ*.

UNDERSTANDING THE MAIN IDEAS

Congruent polygons

Congruent polygons are polygons that have the same size and the same shape. When two polygons are congruent, they can be placed one on top of the other so that their sides and vertices match exactly. Matching sides and vertices of congruent polygons are called *corresponding parts.*

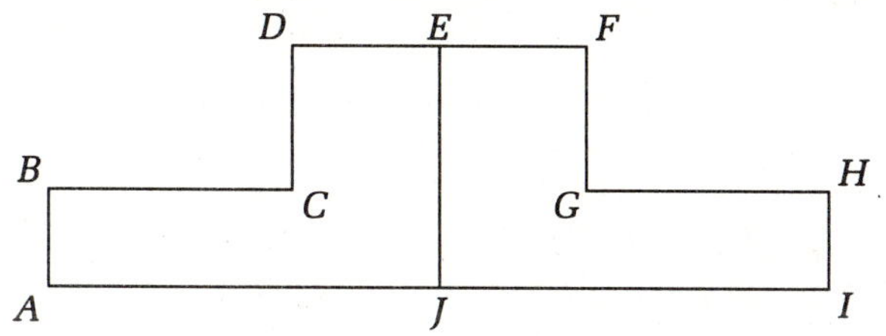

hexagon *ABCDEJ* ≅ hexagon *IHGFEJ*

A corresponds to *I*	side *AB* ≅ side *IH*
B corresponds to *H*	side *BC* ≅ side *HG*
C corresponds to *G*	side *CD* ≅ side *GF*
D corresponds to *F*	side *DE* ≅ side *FE*
E corresponds to *E*	side *EJ* ≅ side *EJ*
J corresponds to *J*	side *JA* ≅ side *JI*

Sample

Write a statement about congruent triangles *ABD* and *CDB*. Then list corresponding sides and corresponding vertices.

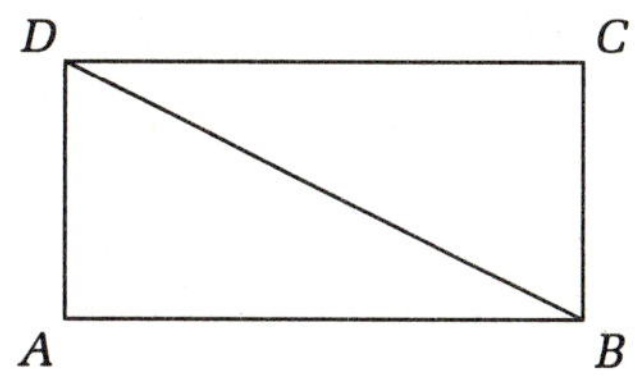

Sample Response

triangle *ABD* ≅ triangle *CDB*

corresponding sides	**corresponding vertices**
sides *AB* and *CD*	vertices *A* and *C*
sides *BD* and *DB*	vertices *B* and *D*
sides *AD* and *CB*	vertices *D* and *B*

For Exercises 1–3, name the congruent polygons. Then list three pairs of corresponding sides.

1.

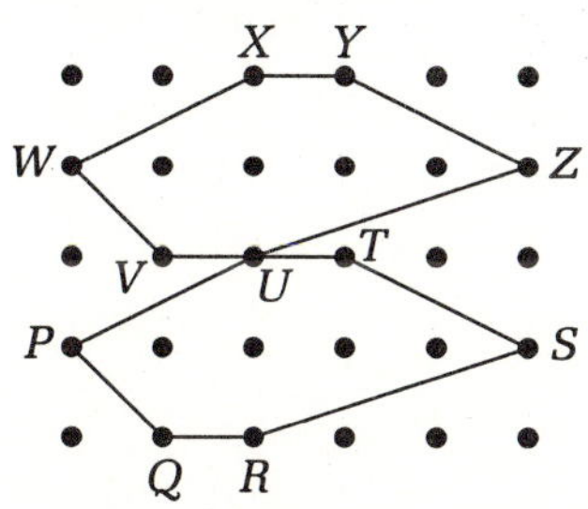

2.

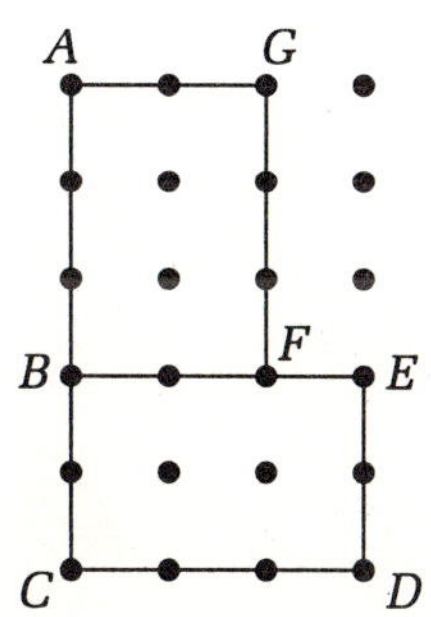

3.

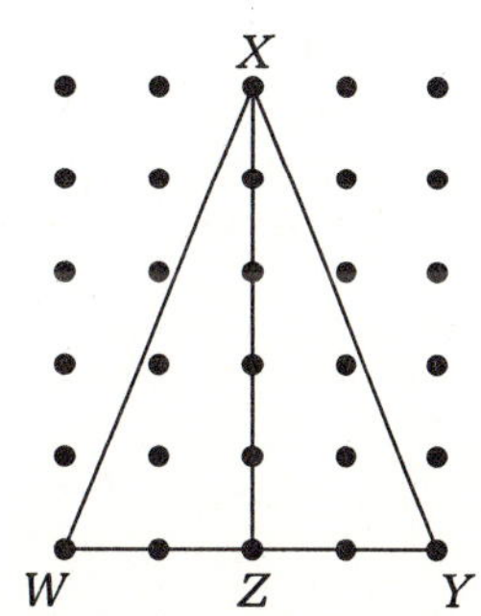

4. ***Open-ended*** Make a logo for a school team by using congruent polygons.

Review PREVIEW

Rewrite without parentheses.

5. $(3x - 2)4$

6. $5(a^2 - 6a - 7)$

Evaluate according to the order of operations.

7. $7 + 10 \div 2$

8. $6 \cdot 8 - 5$

Write a variable expression for the perimeter of each polygon.

9.
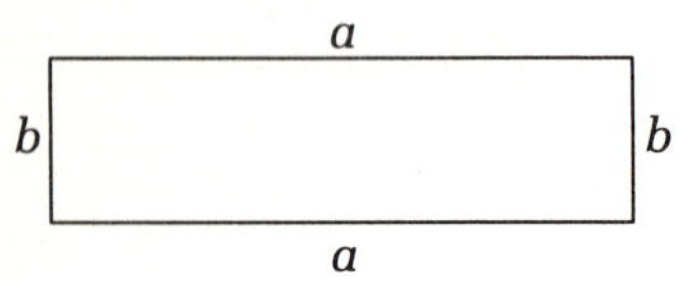

10.
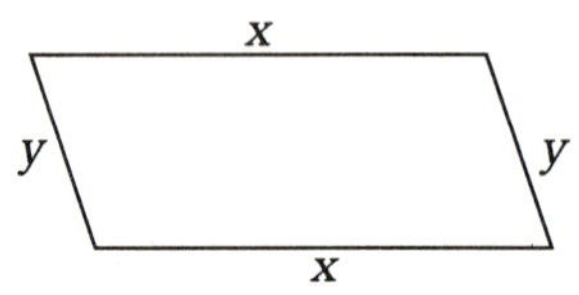

Exploring Quadrilaterals and Symmetry

FOCUS

Build quadrilaterals from congruent triangles and write expressions for their perimeters. Find lines of symmetry. Communicate ideas through writing.

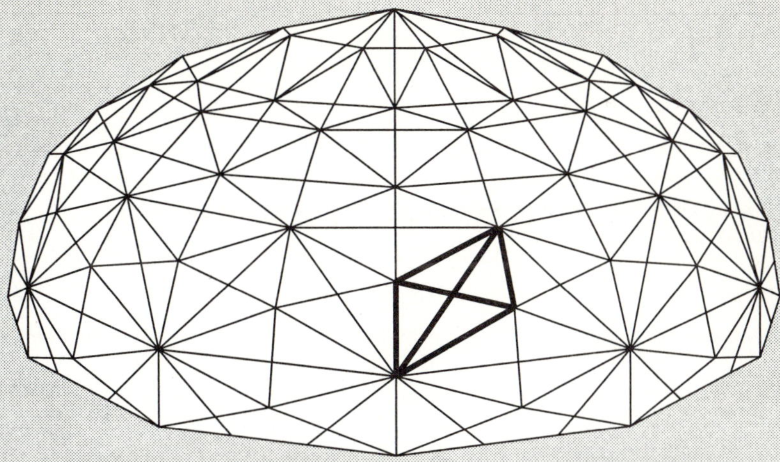

Buckminster Fuller conceived of constructing domes as a network of intersecting triangles. The triangle is the strongest geometric shape. Notice that groups of connected triangles form parallelograms.

KEY TERMS	EXAMPLE / ILLUSTRATION
Equilateral triangle a triangle with all three sides the same length	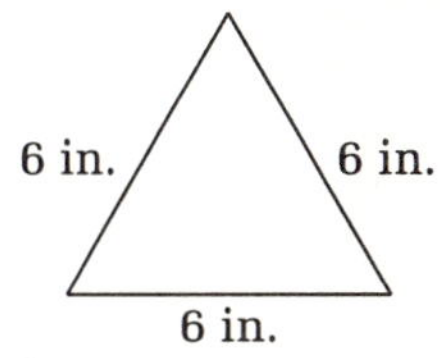
Isosceles triangle a triangle with two sides the same length	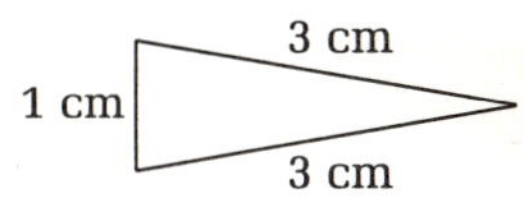
Scalene triangle a triangle with no sides the same length	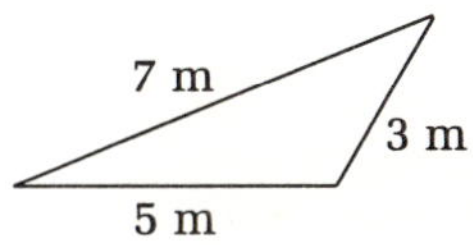
Right triangle a triangle with one right angle	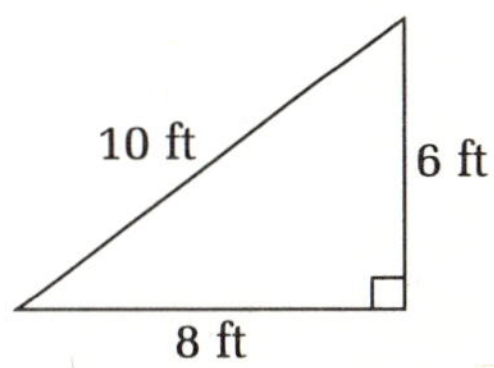
Right angle an angle measuring 90°	In the triangle above, the angle between the 8 ft and 6 ft sides is a right angle.

Term	Example
Perpendicular sides sides of a polygon that meet at a right angle	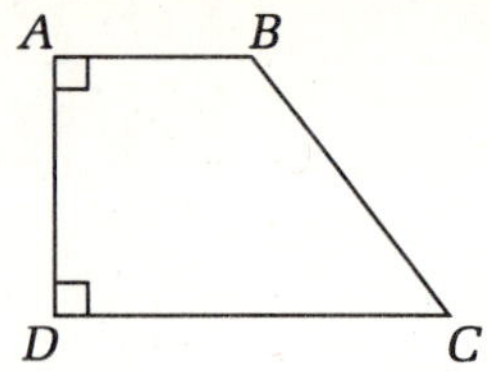 Sides *AB* and *AD* and sides *CD* and *AD* are perpendicular.
Parallel sides sides of a polygon that do not meet (and would never meet if extended indefinitely)	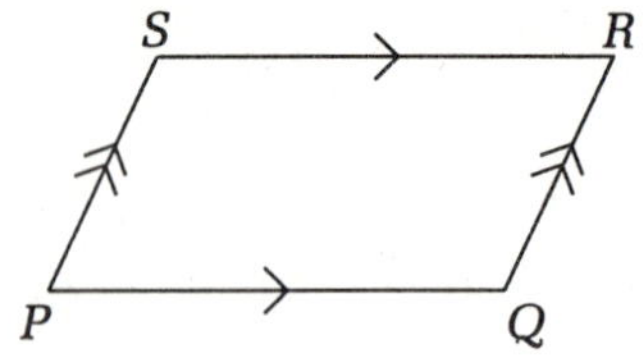 Sides *SR* and *PQ* are parallel. Sides *PS* and *QR* are parallel.
Parallelogram a quadrilateral with both pairs of opposite sides parallel	Quadrilateral *PQRS* above is a parallelogram.
Kite a quadrilateral with two pairs of consecutive sides that are congruent	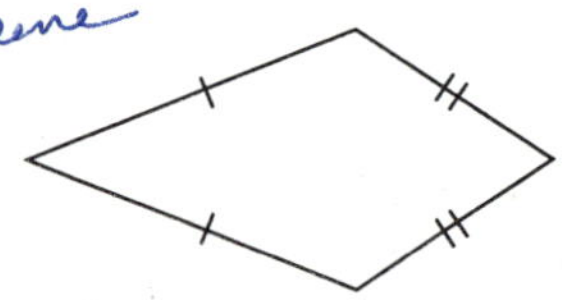
Rhombus a quadrilateral with all four sides congruent	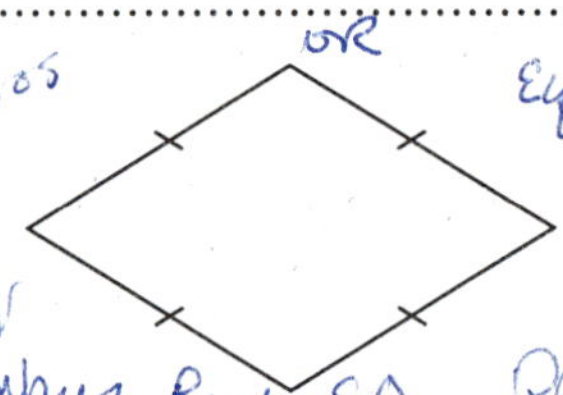
Symmetry a property of some polygons that makes it possible to fold them into matching halves	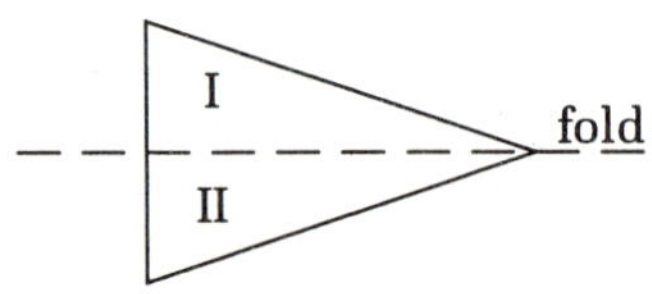 Region I matches region II.
Line of symmetry the "fold line" of a polygon that has symmetry	The fold line in the figure above is the line of symmetry for the triangle.

UNDERSTANDING THE MAIN IDEAS

Describing quadrilaterals

Different types of quadrilaterals are formed when two congruent triangles are positioned so that one pair of congruent sides coincide. Depending on whether the triangles are equilateral, isosceles, or scalene, the resulting quadrilaterals will be rhombuses, kites, parallelograms, rectangles, or squares.

Sample 1

What name best describes the quadrilateral?

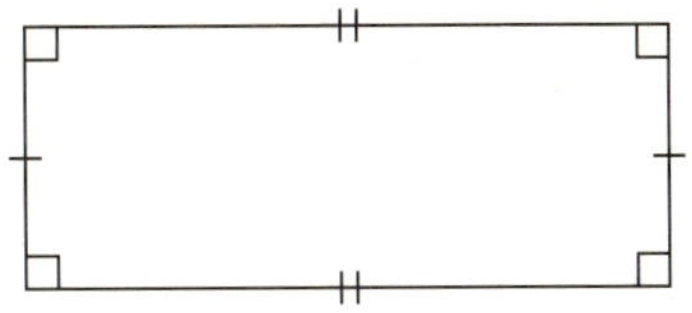

Sample Response

The right angle symbols and tick marks indicate that the quadrilateral is a rectangle.

What name best describes each quadrilateral?

1. 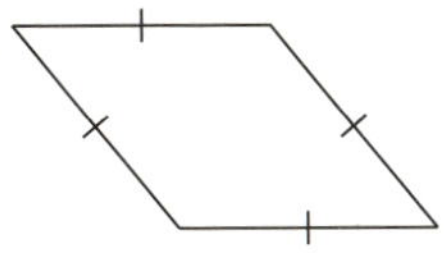

2. 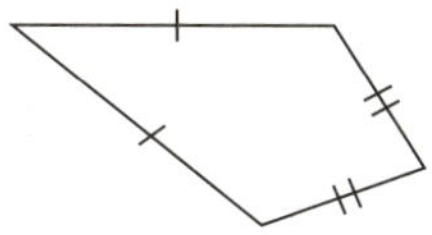

3. 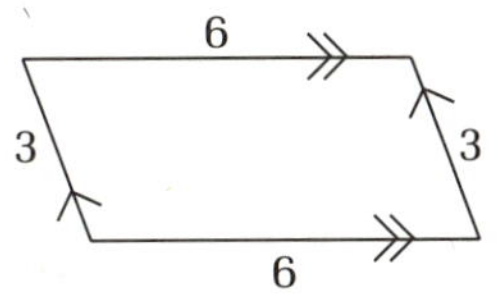

Symmetry

A polygon has symmetry if you can fold a copy of it in such a way that the two parts fit exactly, with one half on top of the other. The fold line is a line of symmetry for the polygon. A figure may have more than one line of symmetry.

Sample 2

Draw the lines of symmetry for each polygon.

a. 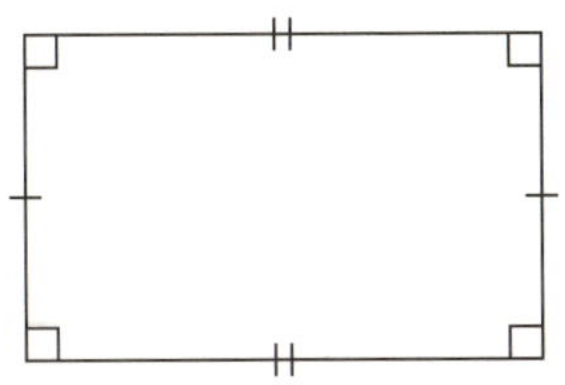

b. 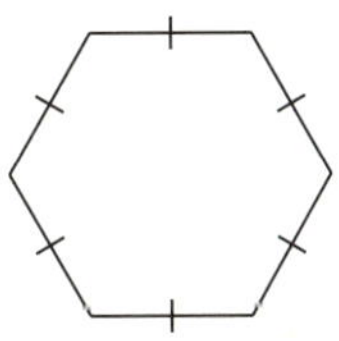

Sample Response

a. 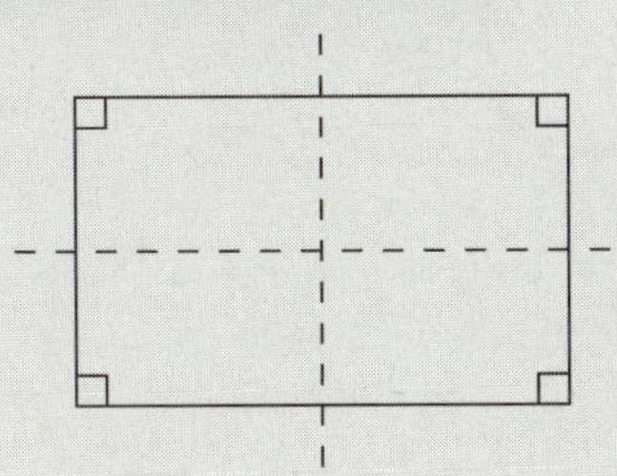

b. 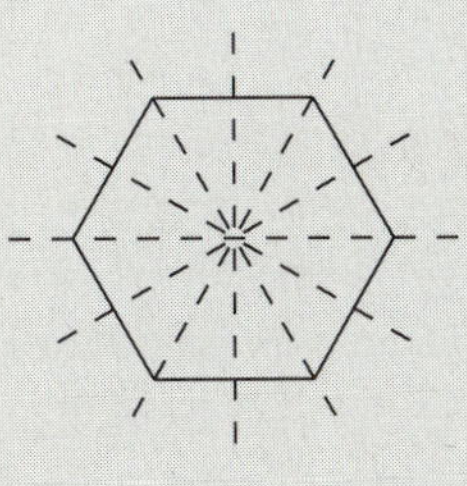

Draw all lines of symmetry for each figure or write *no symmetry*.

4. 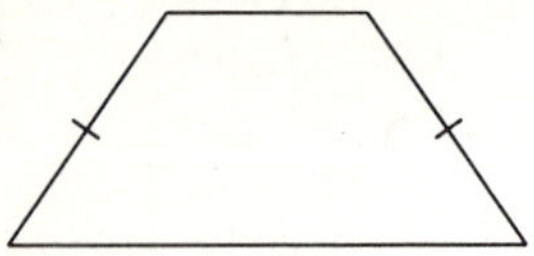

5. 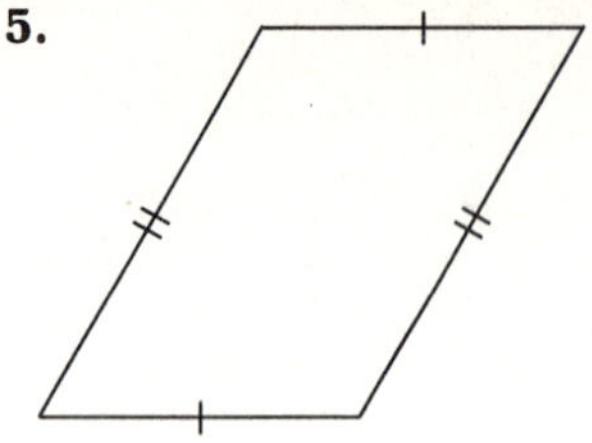

6. 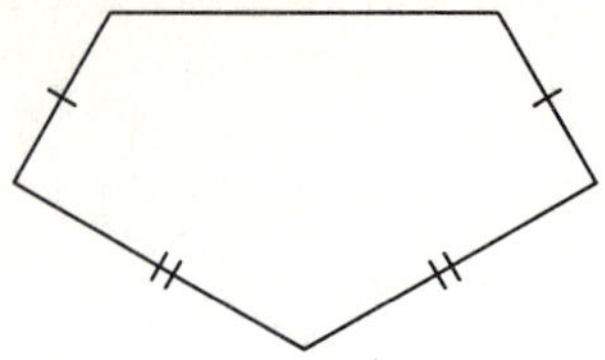

Review PREVIEW

7. Use a triangle and a copy of it to draw a quadrilateral. Label the vertices of the quadrilateral and name some congruent sides.

Combine like terms.

8. $x + y + y + z + x$

9. $p + q + q + q + p$

Write each power of ten as a number and in words.

10. 10^5

11. 10^{12}

12. 10^8

13. 10^1

Area Formulas

FOCUS

Use the formulas for the area of a parallelogram, a triangle, and a trapezoid to solve problem situations.

The two triangular sails on the sailboat are made of canvas. To find the amount of canvas needed to make these sails, you need to know how to find the areas of the two triangles.

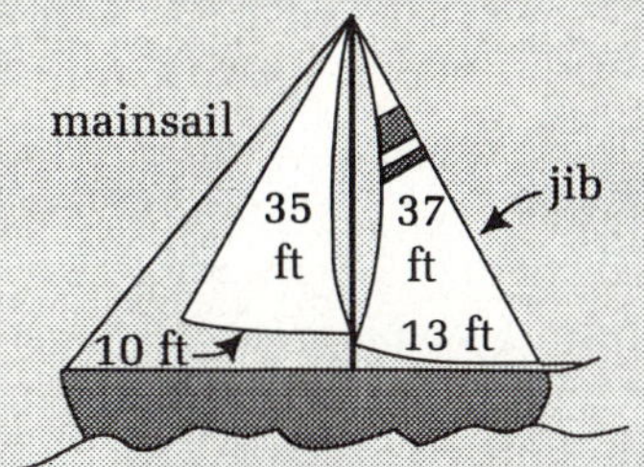

KEY TERMS	EXAMPLE / ILLUSTRATION
Base in a parallelogram or triangle, the length of any side; in a trapezoid, the length of one of the two parallel sides	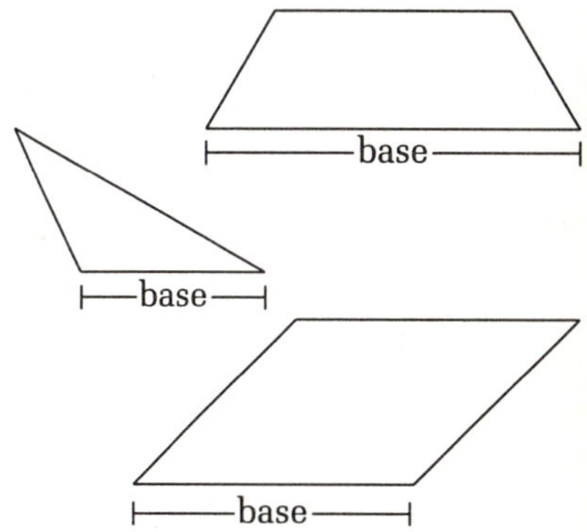
Height in a parallelogram or trapezoid, the perpendicular distance to the base from a point on the opposite side; in a triangle, the perpendicular distance to the base from the opposite vertex	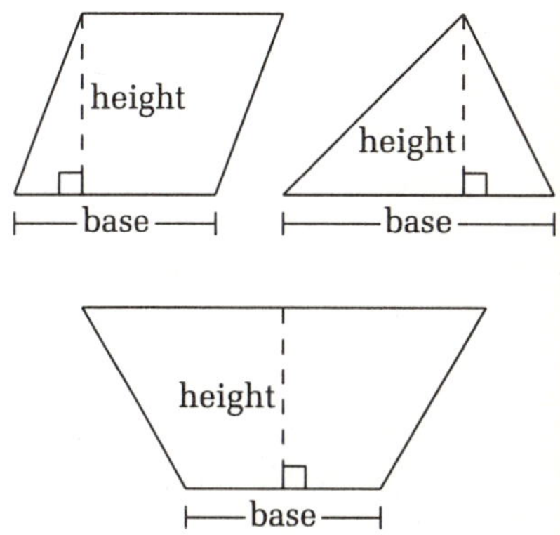

UNDERSTANDING THE MAIN IDEAS

Area of a parallelogram

The area of a parallelogram is found by multiplying its base by its height, that is $A = bh$. The dimensions must be in the same unit and the resulting area is in square units.

Sample 1

Find the area of the parallelogram shown at the right.

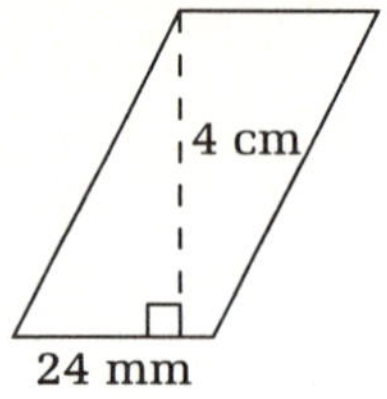

Sample Response

Method 1: Use the formula $A = bh$. First, change 4 cm to 40 mm.

$A = bh$

$A = 24(40)$ ⟵ Substitute 24 for b and 40 for h.

$A = 960$

The area is 960 mm^2.

Method 2: Use the formula $A = bh$. First, change 24 mm to 2.4 cm.

$A = bh$

$A = 2.4(4)$ ⟵ Substitute 2.4 for b and 4 for h.

$A = 9.6$

The area is 9.6 cm^2.

Find the area of each parallelogram.

1.

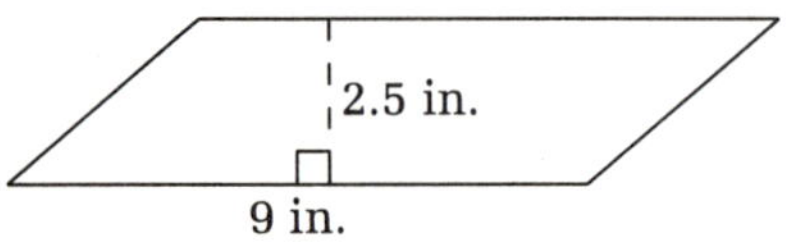

2.

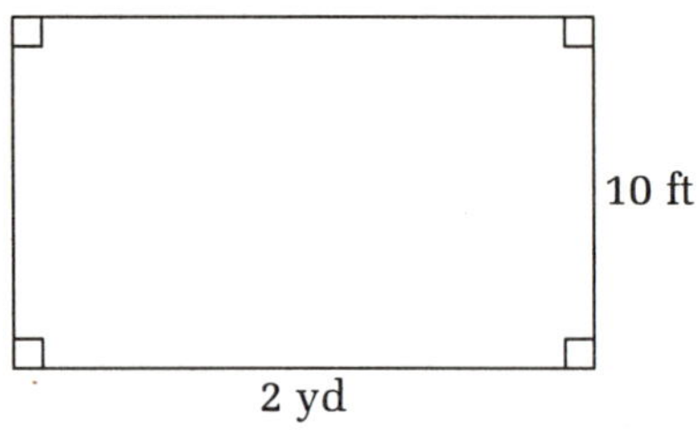

Sample 2

Find the base of a parallelogram with height 6 ft and area 10 yd^2.

Sample Response

Solve the formula $A = bh$ for b.

$A = bh$

$\frac{A}{h} = \frac{bh}{h}$ ⟵ Divide both sides by h.

$\frac{A}{h} = b$

Sample Response continues on the next page.

Before substituting for A and h in the formula $\frac{A}{h} = b$, change 6 ft to 2 yd.

$\frac{10}{2} = b$ ⟵ Substitute 10 for A and 2 for h.

$5 = b$

The base is 5 yd.

3. Find the area of a parallelogram with base 4 m and height 105 cm.
4. Find the height of a parallelogram with area 4 mi^2 and base 2 mi.
5. In Sample 2, if you change the area to 90 ft^2, what is the length of the base in feet?

Area of a triangle

The area of a triangle is found by multiplying $\frac{1}{2}$ by the product of the base and height, that is $A = \frac{1}{2}bh$.

Sample 3

Find the area of the triangle shown at the right.

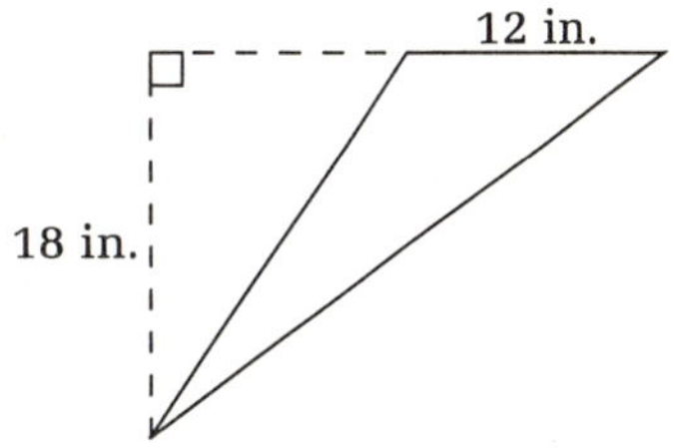

Sample Response

Use the formula $A = \frac{1}{2}bh$.

$A = \frac{1}{2}bh$

$A = \frac{1}{2}(12)(18)$ ⟵ Substitute 12 for b and 18 for h.

$A = 108$

The area is 108 $in.^2$.

Find the area of each triangle.

6.

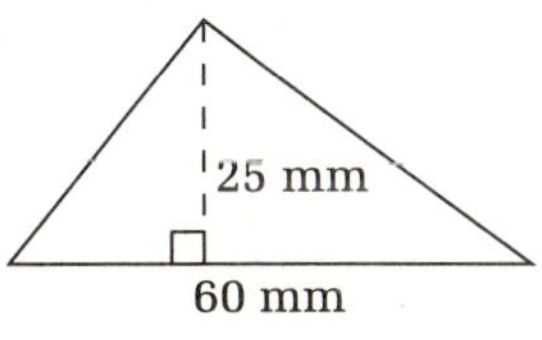

7.

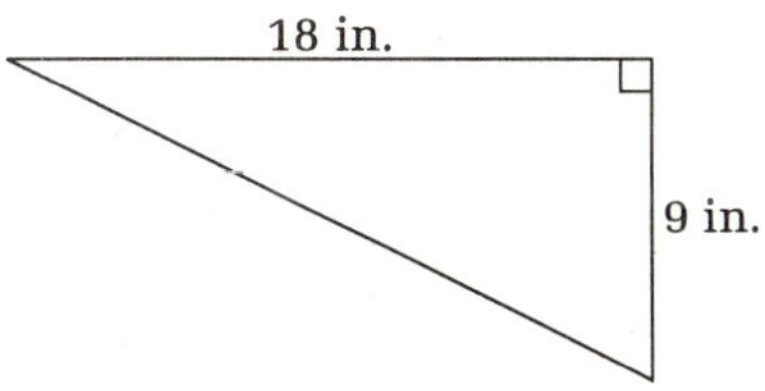

8. Find the area of a triangle with base 2.4 km and height 3 km.

9. Find the height of a triangle with base 10 ft and area 16 ft^2.

Area of a trapezoid

The area of a trapezoid is found by multiplying $\frac{1}{2}$ by the product of the sum of the two bases and the height of the trapezoid, that is

$A = \frac{1}{2}(b_1 + b_2)h.$

Sample 4

Find the area of the trapezoid shown at the right.

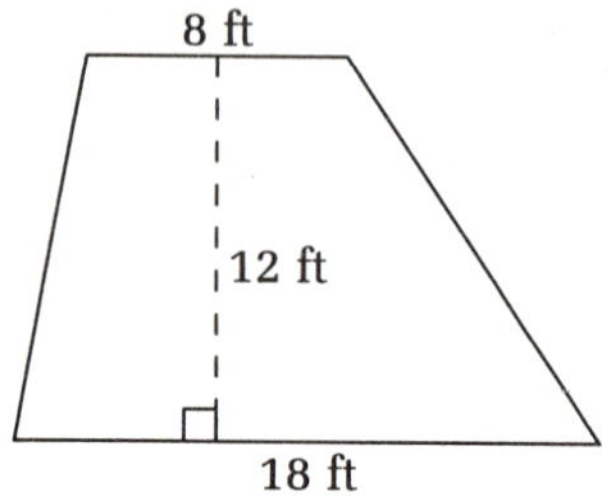

Sample Response

Use the formula $A = \frac{1}{2}(b_1 + b_2)h$.

$A = \frac{1}{2}(b_1 + b_2)h$

$A = \frac{1}{2}(18 + 8)12$ ⟵ Substitute 18 for b_1, 8 for b_2, and 12 for h.

$A = \frac{1}{2}(26)12$

$A = 13(12)$

$A = 156$

The area is 156 ft^2.

Find the area of each trapezoid.

10.

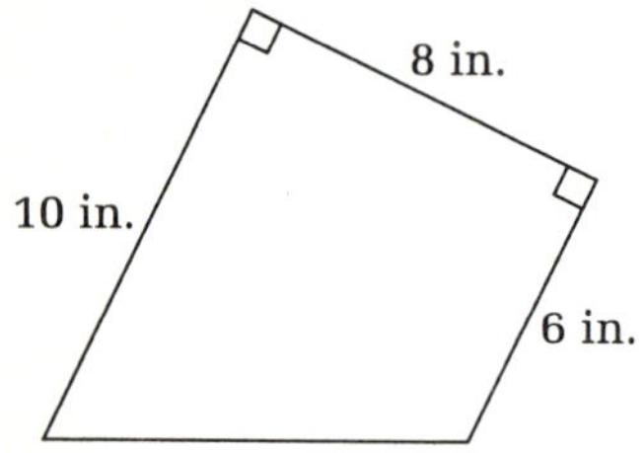

11.

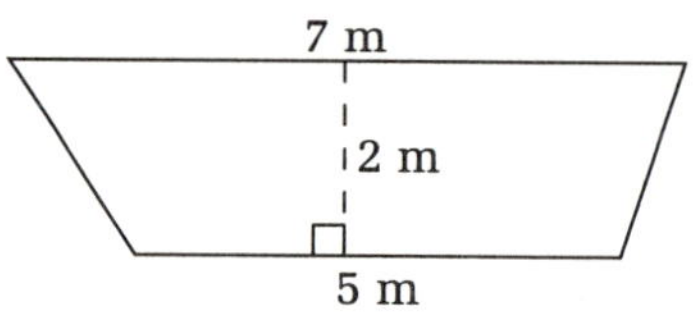

Solving area problems

Area formulas are useful models for many situations. Substituting known values into the appropriate area formula is a useful technique in solving problems involving area.

Sample 5

The Creative Cooking School supplies its students with aprons to wear during classes. The main part of the aprons is trapezoidal in shape. How much fabric will the cooking school need for the main part of 25 aprons?

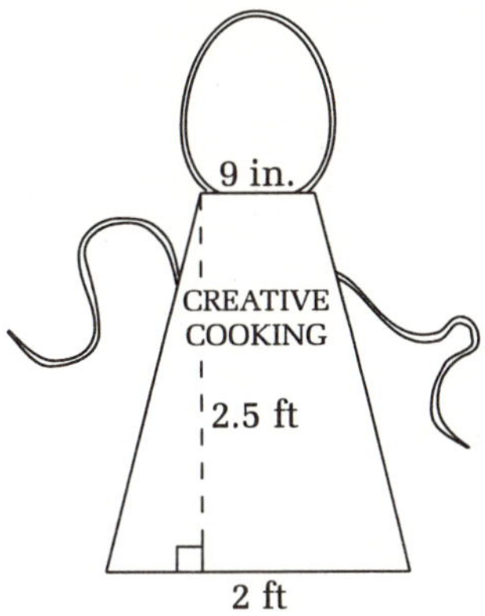

Sample Response

Use the formula $A = \frac{1}{2}(b_1 + b_2)h$, to find the amount of fabric needed for *one* apron. First, change 9 in. to 0.75 ft.

$A = 0.5(2 + 0.75)2.5$ ⟵ Substitute 2 for b_1, 0.75 for b_2, 2.5 for h; use 0.5 for $\frac{1}{2}$.

$A = 0.5(2.75)2.5$

$A = 3.4375$

For one apron, the school will need about 3.5 ft^2 of fabric. To find the amount for 25 aprons, multiply by 25:

$3.5 \cdot 25 = 87.5.$

The school will need about 87.5 ft^2 of fabric for 25 aprons.

Use a formula to solve each problem.

12. The walkway in a shopping mall is paved in stones that repeat the pattern shown at the right. Each stone is 18 in. at the base and has a height of 12 in. What is the area of each paving stone in square inches?

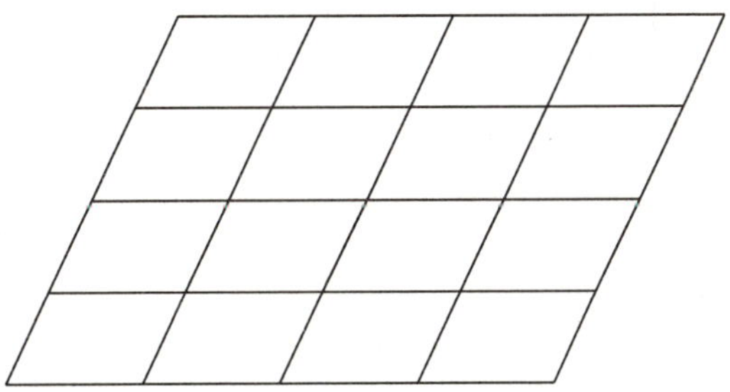

13. Find the amount of canvas needed for the two sails on the boat shown at the right.

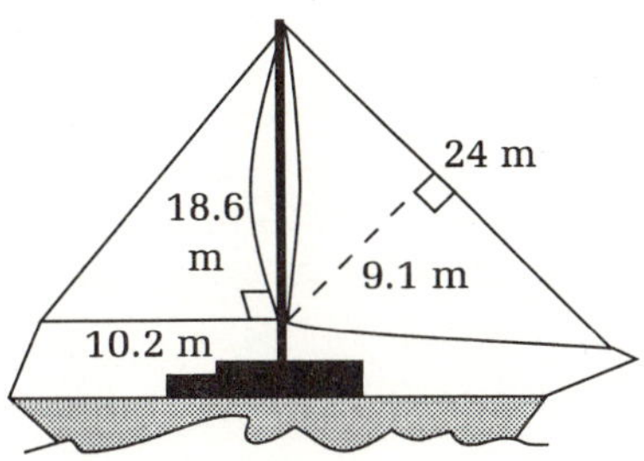

For Exercises 14–16, solve.

14. $\frac{5}{2}x = 30$ **15.** $\frac{3}{4}y - 3 = -9$ **16.** $27 = -\frac{2}{3}x + 3$

17. A quadrilateral has vertices at $A(-6, 2)$, $B(-1, 2)$, $C(1, -2)$, and $D(-8, -2)$. If the quadrilateral is translated 5 units to the right, find the coordinates of the new vertices.

Solve each equation for *y*.

18. $3y - x = 2$ **19.** $4x + 4y = 12$ **20.** $20 = \frac{5}{8}y - 5x$

Circumference and Arc Length

FOCUS

Find the circumference and the length of an arc of a circle.

In one revolution of a car wheel, the car moves forward a distance equal to the circumference of the wheel. For example, if the radius of the wheel is 14.5 in., then the car moves forward about 91 in. during one revolution. During the revolution, a point P on the rim of the wheel takes a curved path, called a cycloid. *The distance traveled by point P for this wheel with 14.5 in. radius is about 116 in.*

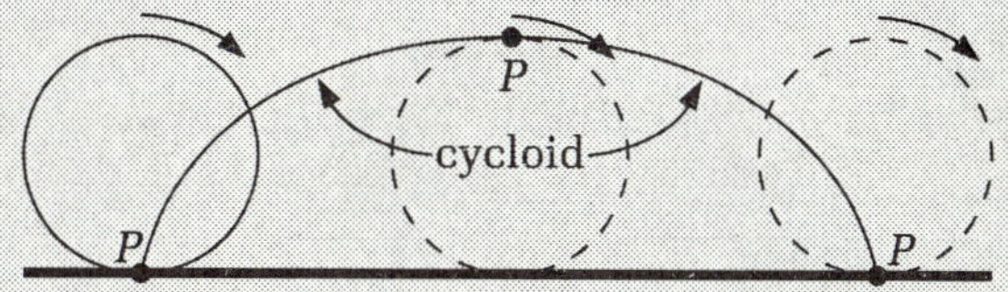

KEY TERMS	EXAMPLE / ILLUSTRATION
Diameter the distance across a circle at its widest part (All diameters pass through the center of the circle.)	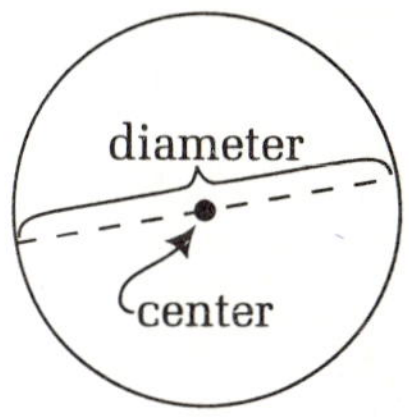
Circumference the distance around the circle	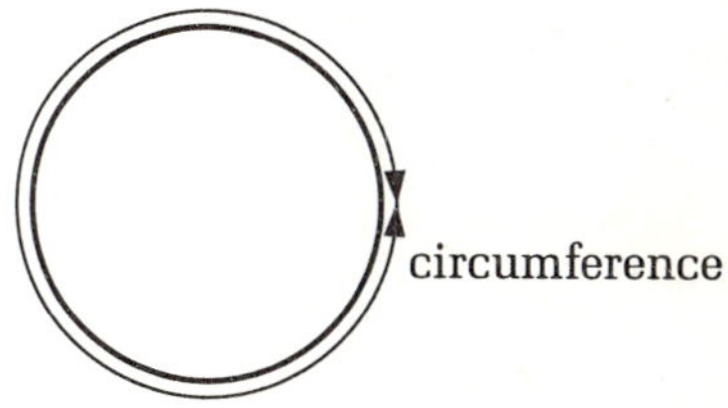
Arc a part of a circle	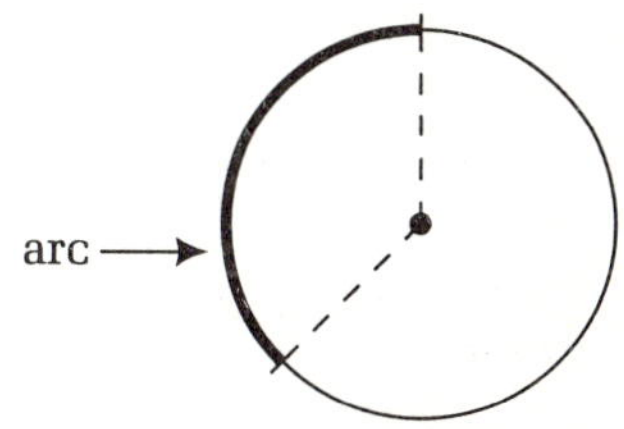
Arc length the length of an arc (The arc length is a fraction of the circumference.)	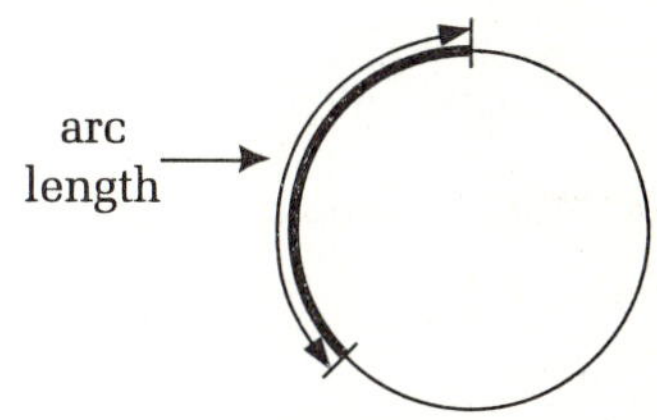
Central angle of an arc the angle, with vertex at the center, determined by the endpoints of an arc	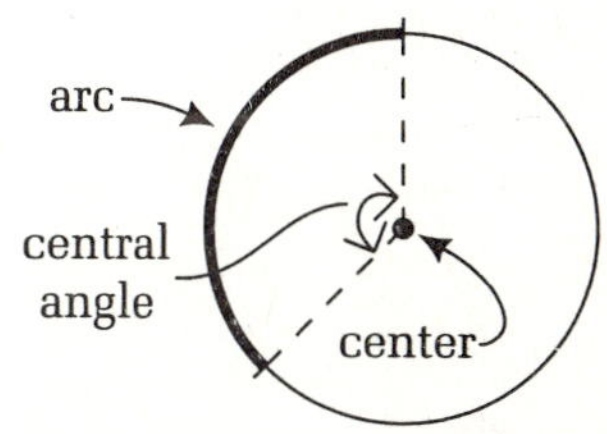

UNDERSTANDING THE MAIN IDEAS

The ratio pi

In all circles, the ratio of the circumference to the diameter is a constant, represented by the Greek letter pi (π). This ratio cannot be represented exactly as a fraction or decimal. A good decimal approximation for the value of π is 3.14. To find the circumference, C, of a circle, use the formula $C = \pi d$, where d represents the diameter of the circle.

Sample 1

Find the diameter of a circular water tower that is 53 m around.

Sample Response

Use the formula for circumference, $C = \pi d$.

$C = \pi d$

$53 \approx (3.14)d$ ⟵ The "distance around" the tower, 53 m, is the circumference.

$\frac{53}{3.14} \approx \frac{3.14d}{3.14}$

$16.9 \approx d$ ⟵ This answer is given to the nearest tenth.

The diameter of the tower is approximately 16.9 meters.

1. What is the radius of a circle with circumference 11 ft?
2. You want to plant 100 tulip bulbs 6 in. apart in a circle. To the nearest foot, what is the diameter of the plot that the tulips will border?
3. How much fencing will be needed to enclose a circular plot measuring 15 ft in diameter?

Arc length

Arc length is a fraction of the circumference of the circle. To find the length of an arc, you can use this proportion:

$$\frac{\text{arc length}}{\text{circumference}} = \frac{\text{measure of central angle of arc}}{360^\circ}$$

Sample 2

Find the arc length.

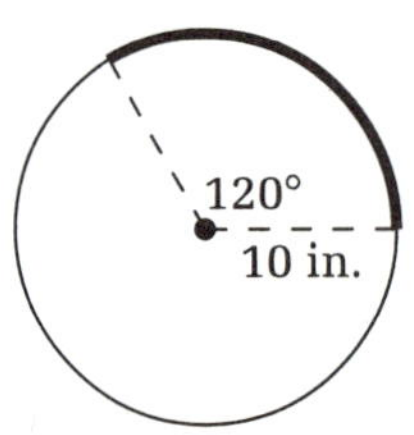

Sample Response

Use a proportion. Let l stand for the arc length.

$$\frac{l}{C} = \frac{\text{measure of central angle}}{360°}$$

Since the radius of the circle is 10 in., the diameter is 20 in. So, the circumference is $C = \pi d \approx 3.14(20)$, or about 62.8 in.

$$\frac{l}{62.8} = \frac{120}{360}$$

$$\frac{l}{62.8} = \frac{1}{3} \longleftarrow \text{Write the ratio in lowest terms.}$$

$$62.8\left(\frac{l}{6.28}\right) = 62.8\left(\frac{1}{3}\right)$$

$$l \approx 20.9$$

The arc length is approximately 20.9 in.

Find each arc length. Give answers to the nearest tenth.

4.

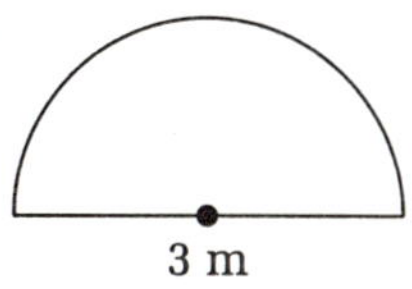

5.

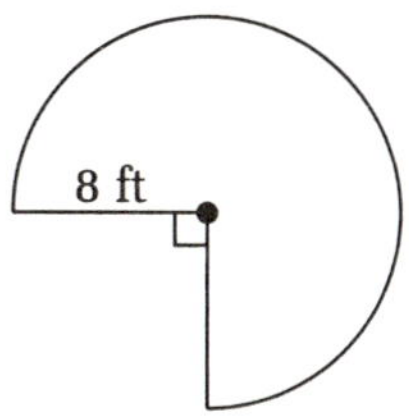

6.

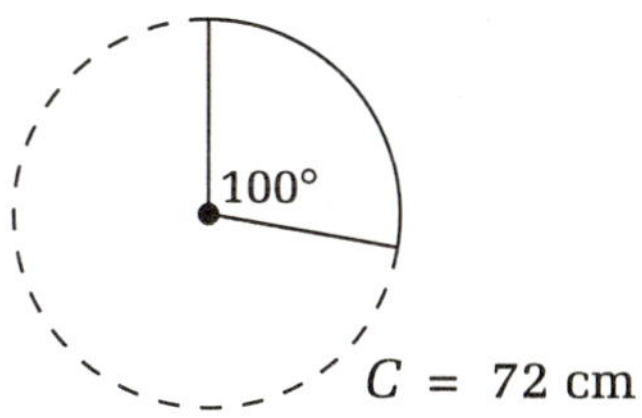

7. The clock on the city hall has a minute hand one meter long. A bug sits on the tip of the minute hand for two minutes. What distance does it ride? Give your answer to the nearest centimeter.

8. Find the distance around the toy racetrack shown. Give your answer to the nearest tenth.

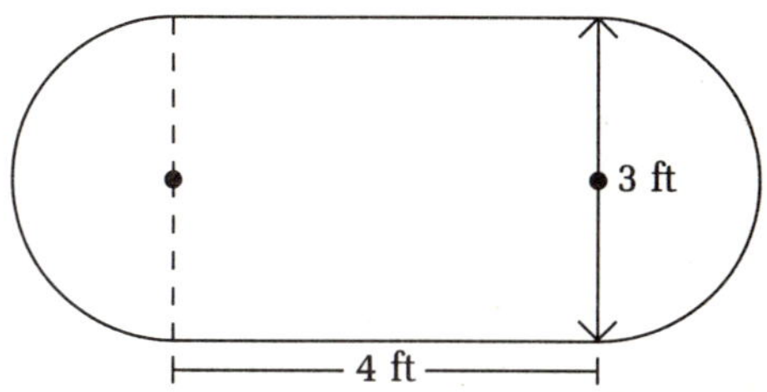

Review PREVIEW

Draw a line through the points and find its slope.

9. (0, 0) and (3, –2)

10. (4, 2) and (–3, 8)

11. Draw a box-and-whisker plot for the times customers waited in line for service at a bank.

Customer waiting times (minutes)				
7	4	6	12	8
8	5	2	5	7
9	10	6	15	3

Evaluate each expression for the given value of the variable.

12. $-6x$ when $x = -2$

13. $0.125x$ when $x = 160$

Areas of Circles and Sectors

FOCUS

Use the formulas for the area of a circle and find the area of a sector.

Ninety-degree sectors are used in this traditional American quilt pattern called "Around the World."

KEY TERMS

KEY TERMS	EXAMPLE / ILLUSTRATION
Sector the region of a circle formed by a central angle and its arc	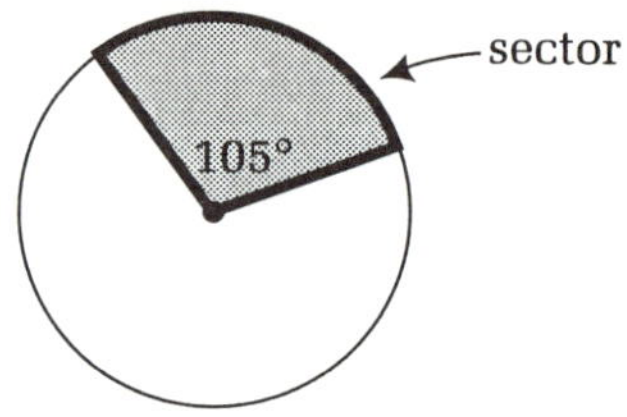

UNDERSTANDING THE MAIN IDEAS

Areas of circles

For all circles, the ratio of the area of the circle to the *square* of the radius of the circle is π. To find the area of a circle, use the formula $A = \pi r^2$.

Sample 1

Find the area of the circle shown at the right. Give your answer to the nearest tenth.

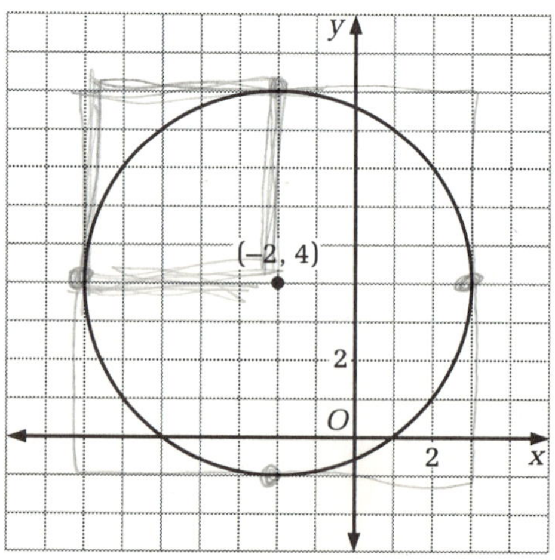

Sample Response

To use the formula $A = \pi r^2$ to find the area of the circle, you need to know the length of the radius of the circle. You can determine this information from the graph. Since the circle passes through $(-7, 4)$, the radius of the circle is $-2 - (-7) = 5$.

Substitute $r = 5$ and $\pi \approx 3.14$ into the formula $A = \pi r^2$:

$A \approx (3.14)(5^2)$
$A \approx 3.14(25)$
$A \approx 78.5$

The area of the circle is approximately 78.5 square units.

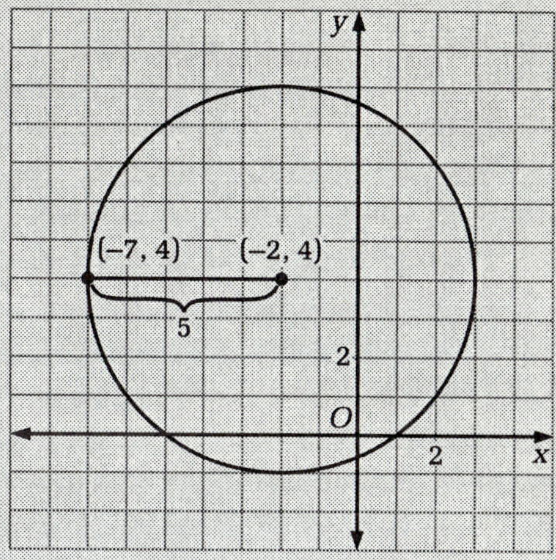

For Exercises 1–3, find the unknown measurement for each circle. Give your answers to the nearest tenth.

1. radius = 6 in.
   area = ___?___
2. area = 201 km^2
   radius = ___?___
3. area = 18.1 m^2
   diameter = ___?___
4. Graph the circle with center at $(3, -2)$ and area 16π square units.

Area of a sector

To find the area of a sector, use this proportion:

$$\frac{\text{area of a sector}}{\text{area of the circle}} = \frac{\text{measure of the central angle}}{360^\circ}.$$

Sample 2

Notice that the "Around the World" quilt pattern on the first page of this study lesson is made with 15 rotations and translations of the square shown at the right.

If the sides of the square measure 5 in. and the radius of the shaded sector is 4 in., find the area of the light-colored cloth used in this square.

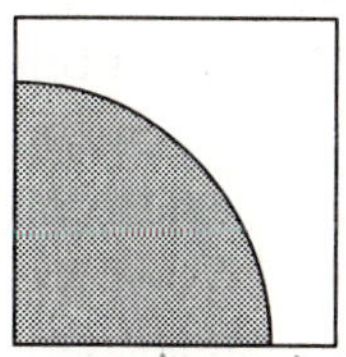

Sample Response

Break the problem into parts: (1) Find the area of the square. (2) Find the area of the dark-colored sector. (3) Subtract the area of the sector from that of the square.

(1) Area of square $= 5^2 = 25$ in.^{2}

Sample Response continues on the next page.

(2) To find the area of the sector, use the proportion

$$\frac{\text{area of sector}}{\text{area of circle}} = \frac{\text{measure of the central angle}}{360°}.$$

Since the central angle of the sector is one corner of the square, its measure is 90°. Recall that the radius of the sector is 4 in., so the area of the circle is $\pi(4^2) \approx 50.24$ in.^{2}. Letting A = area of the sector:

$$\frac{A}{50.24} = \frac{90}{360}$$

$$\frac{A}{50.24} = \frac{1}{4}$$

$$A = \frac{50.24}{4}$$

$$A = 12.56 \text{ in.}^2$$

(3) Now subtract to find the area of the light-colored cloth:

$$\begin{aligned}\text{area of light cloth} &= \text{area of square} - \text{area of sector}\\ &= 25 - 12.56\\ &= 12.44\end{aligned}$$

The area of the light-colored cloth is approximately 12.44 in.^{2}.

In Exercises 5–7, find the area of each sector to the nearest tenth.

5.

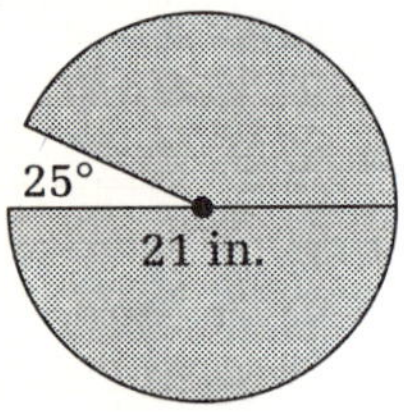

6.

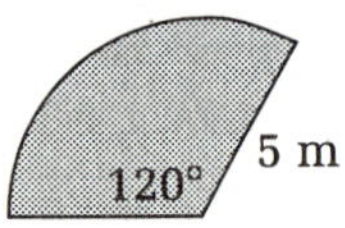

7.

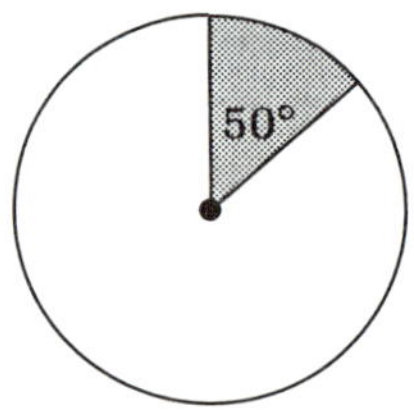

The area of the circle is 121π km^2.

8. Which is larger: a slice of a 12-in. pizza which has been cut into 6 equal pieces, or a slice of a 14-in. pizza which has been cut into 8 equal pieces? Both pizzas have the same thickness.

Review PREVIEW

For Exercises 9 and 10, use conversion factors.

9. Convert 0.06 mi/h to feet per second.

10. Convert 5¢/in. to dollars per yard.

11. The graph at the right shows a rotation around the origin. What is the direction of the rotation and the amount of turn?

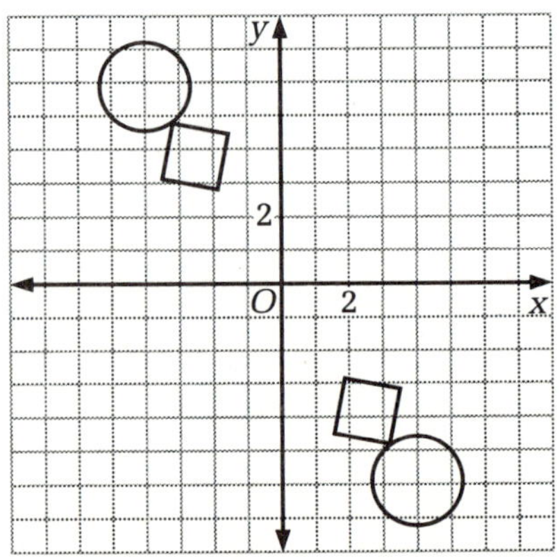

Solve.

12. $-4 = 0.25x + 6$

13. $\frac{x}{12} - 3 = 42$

14. $5 = -3x - 16$

Surface Area of Space Figures

FOCUS

Find the surface area of prisms, cylinders, and pyramids.

A store manager is planning to shrink-wrap packages of three VCR tapes for a special sale. The manager wants to know how much plastic material is needed to shrink-wrap 100 packages of tapes. The manager will need to find the surface area of a stack of three tapes in order to know how much plastic is needed per package.

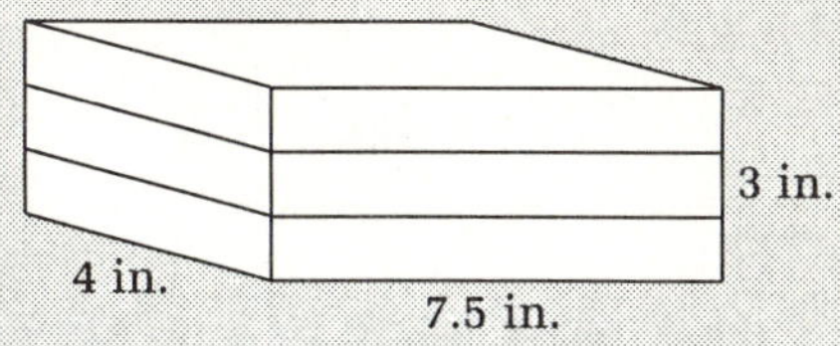

KEY TERMS	EXAMPLE / ILLUSTRATION
Prism a space figure with two congruent, parallel polygons called bases (The other sides of the prism are called faces.)	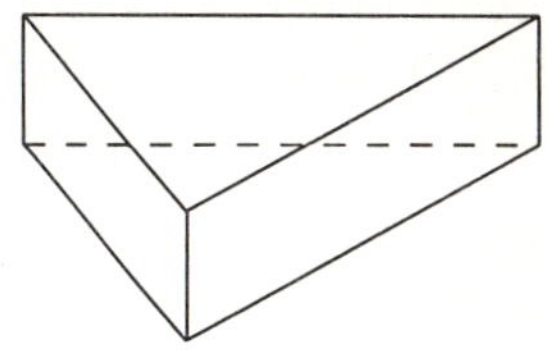
Bases of a prism the two congruent, parallel polygonal sides of a prism	The bases of the prism shown above are the two congruent triangular sides.
Faces of a prism the sides of a prism that are not bases	The faces of the prism shown above are the three rectangular sides.
Cylinder a space figure with a curved surface and two congruent, parallel bases that are circles	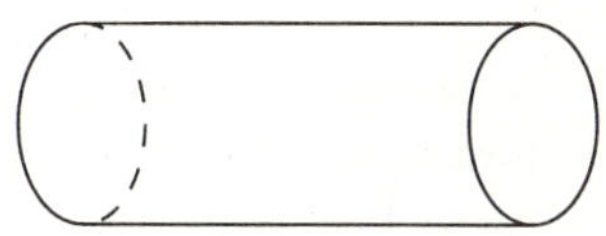
Pyramid a space figure with a polygon as its one base and triangular faces	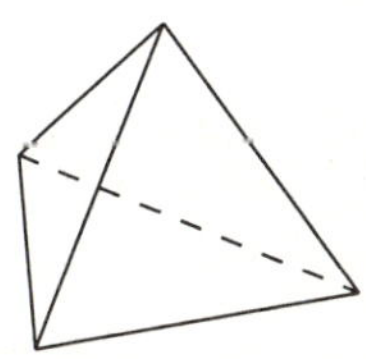
Regular pyramid a pyramid whose base is a regular polygon and whose other faces are congruent isosceles triangles	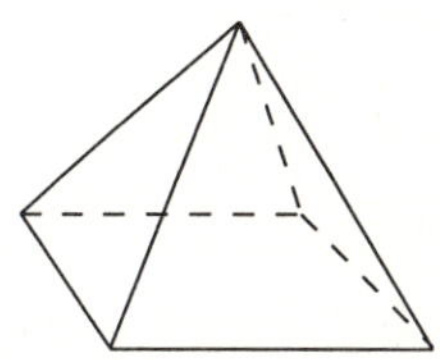
Height of a regular pyramid the length of the perpendicular segment from the vertex to the center of the base	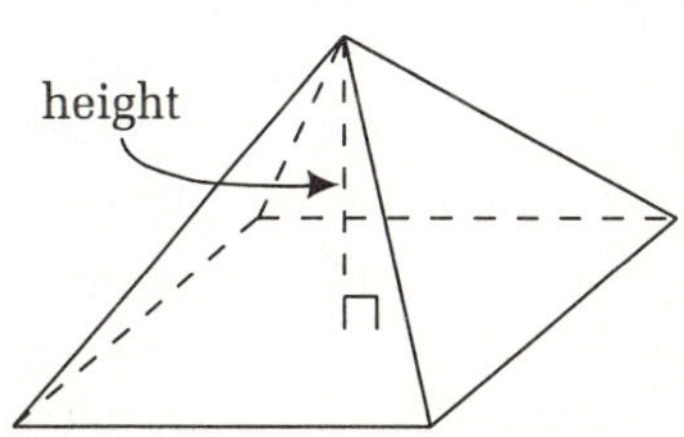

Slant height of a regular pyramid
the height of any one of the congruent triangular faces of a regular pyramid

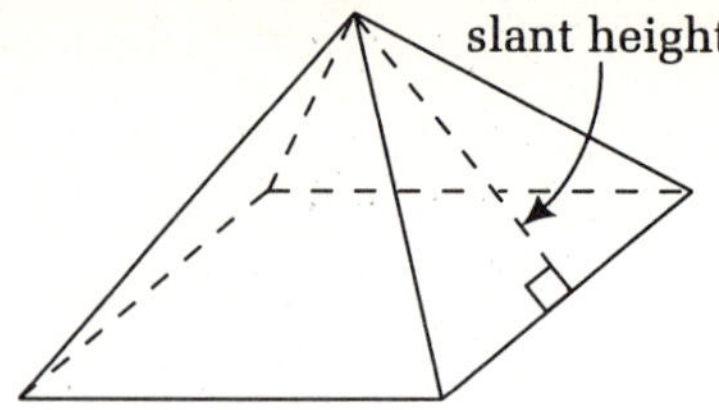

UNDERSTANDING THE MAIN IDEAS

Surface area of a prism

The surface area of a prism is found by using the formula

Surface Area (S.A.) = area of two bases + areas of faces.

Since the bases are congruent, both bases have the same area. If the bases are triangles, the prism has three rectangular faces; if the bases are rectangles, the prism has four rectangular faces; and so on. Surface area is measured in square units.

Sample 1

All the sides of this trapezoidal prism are to be covered in red felt. How many square yards of felt are needed?

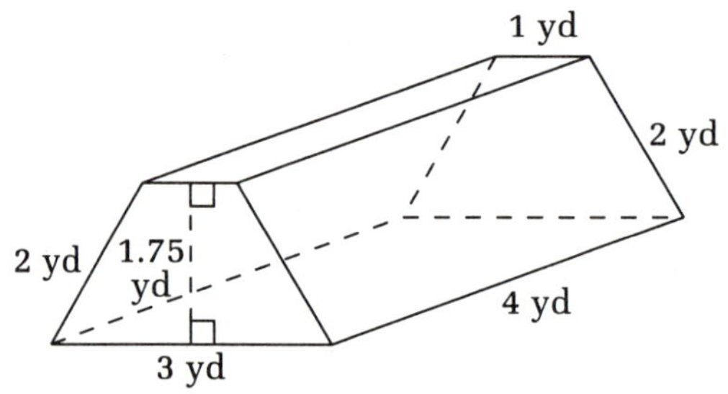

Sample Response

Use the formula S.A. = area of two bases + area of faces.

Step 1 Find the area of the two bases.

The bases are congruent trapezoids with bases 1 yd and 3 yd long, and height 1.75 yd. Use the formula for the area of a trapezoid to find the area of each congruent base.

$A = \frac{1}{2}(b_1 + b_2)h$

$= \frac{1}{2}(1 + 3)1.75$ ⟵ Substitute for b_1, b_2, and h.

$= 3.5$

Now, multiply by 2 to find the area of both bases: $2 \times 3.5 = 7$ yd^2.

Step 2 Find the area of the faces.

There are four rectangular faces.

Two of the faces are 4 yd-by-2 yd rectangles: $A = 2(4 \times 2) = 16$ yd^2.
One of the other faces is a 1 yd-by-4 yd rectangle: $A = 1 \times 4 = 4$ yd^2.
The remaining face is a 3 yd-by-4 yd rectangle: $A = 3 \times 4 = 12$ yd^2.

Now, add to find the total area of all four faces: $16 + 4 + 12 = 32$ yd^2.

Sample Response continues on the next page.

Step 3 Find the total area.

Substitute the results from steps 1 and 2 into the formula

S.A. = area of two bases + area of faces.

S.A. = 7 + 32 = 39 yd^2

A total of 39 yd^2 of felt are needed to cover the prism.

Identify each type of prism and find its surface area.

1.

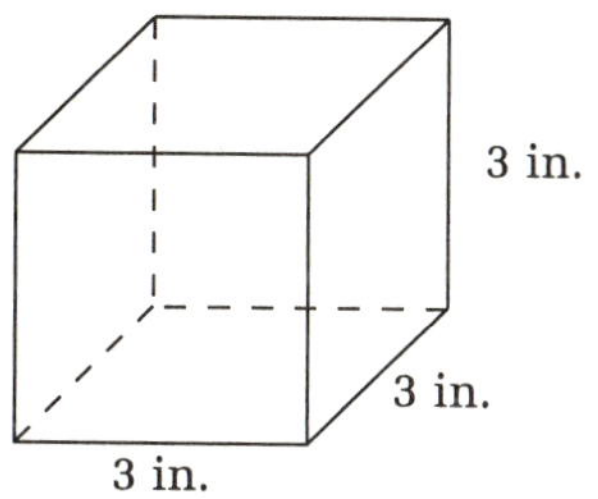

2.

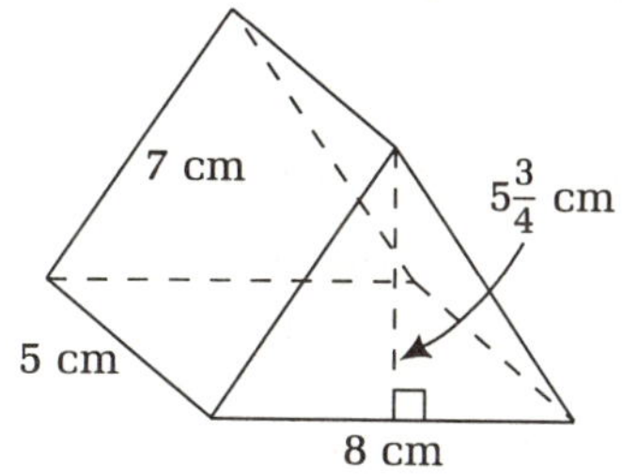

Surface area of a cylinder

The surface area of a cylinder is found by using the formula

Surface Area = area of two circular bases + area of curved surface

S.A. = $2\pi r^2 + 2\pi rh$.

In the formula, the expression for the area of the curved surface is found by noting that the curved surface could be "flattened out" to form a rectangle with length equal to the circumference of the circular base, $2\pi r$, and width equal to the height of the cylinder, h. As for prisms, the surface area of a cylinder is measured in square units.

Sample 2

Find the surface area of the half cylinder shown below.

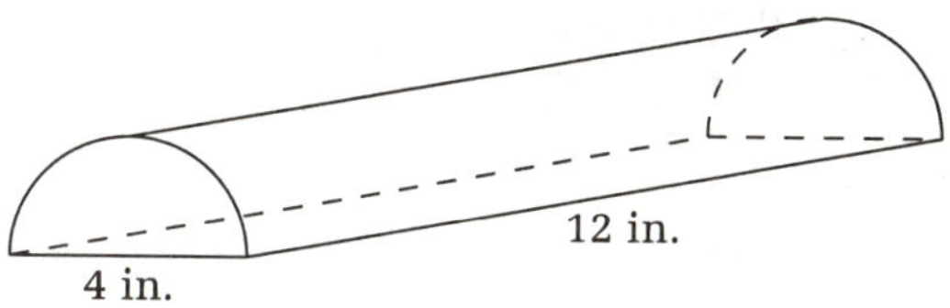

Sample Response

Step 1 Find the area of the curved surface and the two half-circle bases.

Use the formula for the surface area of a cylinder. Because you are finding the surface area of a half cylinder, divide both sides of the formula by 2.

$$\frac{\text{S.A.}}{2} = \frac{2\pi r^2 + 2\pi rh}{2}$$

$$= \pi r^2 + \pi rh$$

Since the diameter of the half cylinder is 4 in., the radius is $4 \div 2 = 2$ in.

$$\frac{\text{S.A.}}{2} = \pi(2)^2 + \pi(2)(12)$$ ⟵ Substitute 2 for r and 12 for h.

$$= 4\pi + 24\pi$$

$$= 28\pi$$

$$\approx 88 \text{ in.}^2$$

Sample Response continues on the next page.

Step 2 Find the area of the flat surface.
The flat surface is a 4 in.-by-12 in. rectangle.

$A = 4 \times 12$
$= 48 \text{ in.}^2$

Step 3 Find the total area.
Add the areas found in steps 1 and 2.

S.A. $\approx 88 + 48 \approx 136 \text{ in.}^2$

The surface area of the half cylinder is about 136 in.^{2}.

Find the surface area of each cylinder or half cylinder.

3.

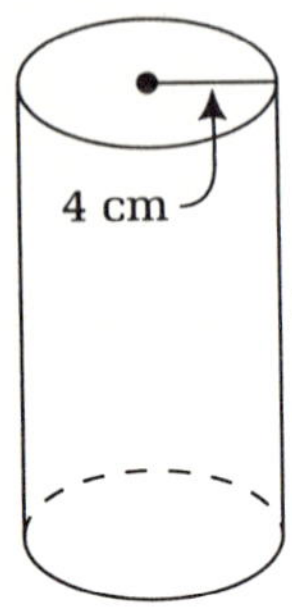

4.

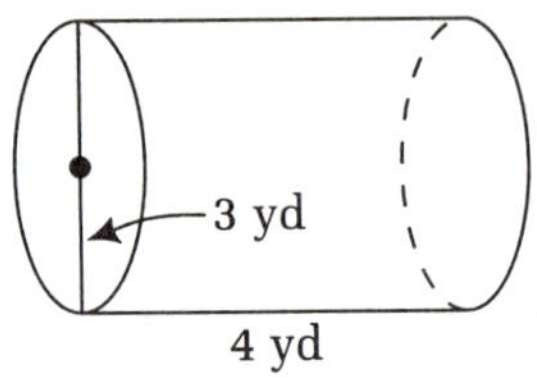

5.

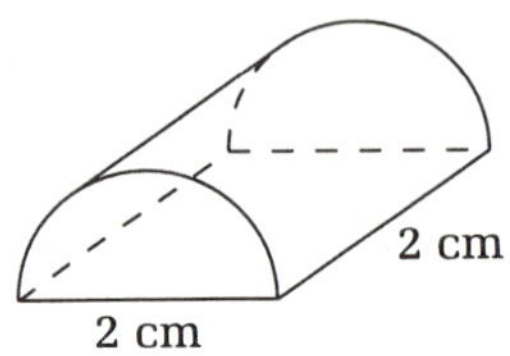

Surface area of a pyramid

The surface area of a pyramid is found by using the formula

S.A. = area of base + area of faces.

In a regular pyramid, all the faces are congruent isosceles triangles, so you need only find the area of one face and then multiply by the number of faces to find the total area of the faces. As for prisms and cylinders, surface area is measured in square units.

Sample 3

Find the surface area of this regular triangular pyramid.

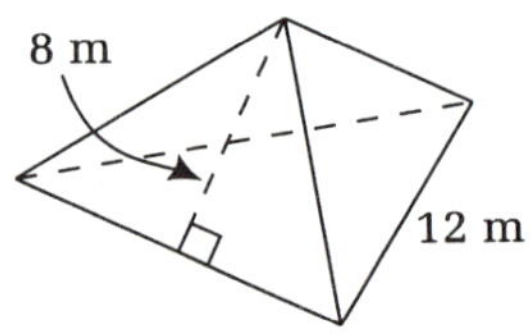

Sample Response

Use the formula S.A. = area of base + area of faces.

Step 1 Find the area of the base.
Since the pyramid is a regular triangular pyramid, the base is an equilateral triangle. First, you need to find the height of the base. (*Note:* The height of an equilateral triangle divides the side to which it is drawn in half, as shown in the figure.)

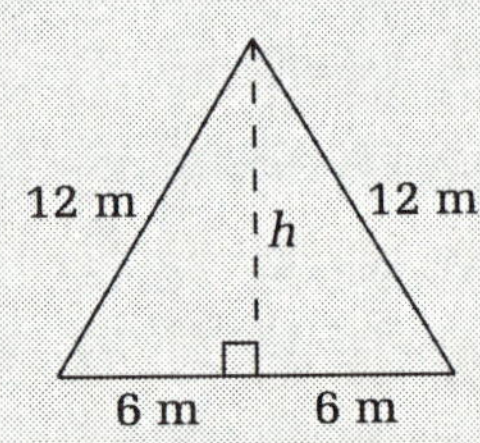

Sample Response continues on the next page.

By the Pythagorean theorem,

$$6^2 + h^2 = 12^2$$
$$36 + h^2 = 144$$
$$h^2 = 108$$
$$h \approx 10.4 \text{ m}$$

Now use the formula for the area of a triangle to find the area of the base.

$$A = \frac{1}{2}bh$$
$$\approx \frac{1}{2}(12)(10.4) \longleftarrow \text{Substitute 12 for } b \text{ and 10.4 for } h.$$
$$\approx 62.4 \text{ m}^2$$

Step 2 Find the area of the faces.

The base and height of each of the three congruent triangular faces are 12 m and 8 m, respectively. Use the formula for the area of a triangle to find the area of each congruent face.

$$A = \frac{1}{2}bh$$
$$= \frac{1}{2}(12)(8) \longleftarrow \text{Substitute 12 for } b \text{ and 8 for } h.$$
$$= 48 \text{ m}^2$$

Now multiply by 3 to find the area of all three faces.

$A = 3 \times 48 = 144 \text{ m}^2$

Step 3 Find the total area.

Add the areas found in steps 1 and 2.

S.A. $\approx 62.4 + 144 \approx 206.4 \text{ m}^2$

The surface area of the regular triangular pyramid is about 206.4 m^2.

Find the surface area of each regular pyramid.

6.

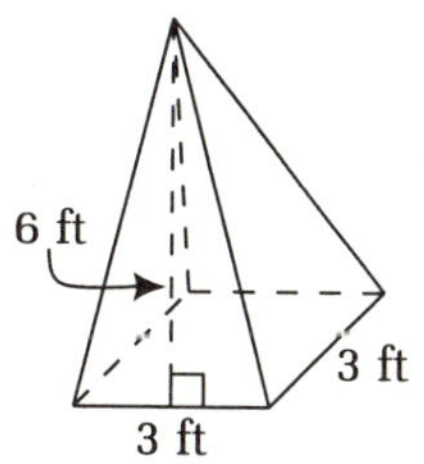

7.

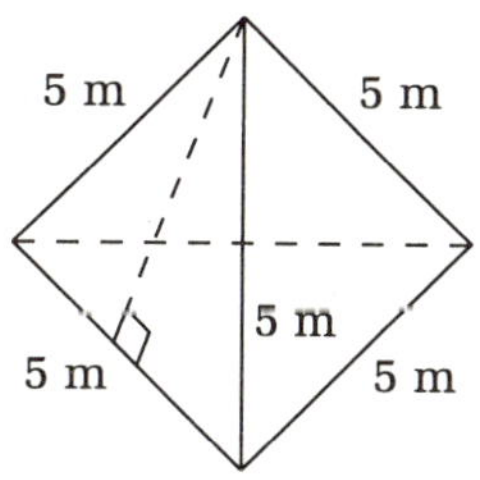

Review **PREVIEW**

For Exercises 8–10, suppose a point on $\overline{RT}$ is picked at random. Find the probability that the point is on each segment named.

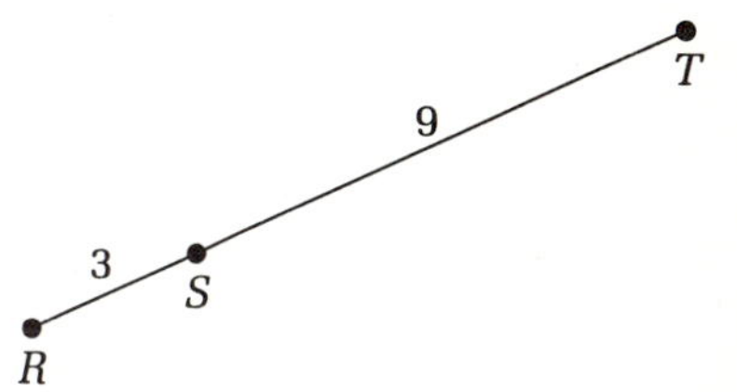

8. $\overline{ST}$ **9.** $\overline{RS}$ **10.** $\overline{RT}$

11. Sketch a parallelogram. To find the area of your parallelogram, what dimensions do you need to know?

Find the volume of each box.

12. a box with length 12 in., width 7 in., and height 10 in.

13. a box with length 6 cm, width 6 cm, and height 5 cm

Volumes of Prisms and Cylinders

FOCUS

Find the volume of prisms and cylinders.

The volume of the prism shown at the right is 1000 cm^3, which is equivalent to 1 liter. How could you determine the diameter of a 1-liter cylindrical container that is 10 cm tall?

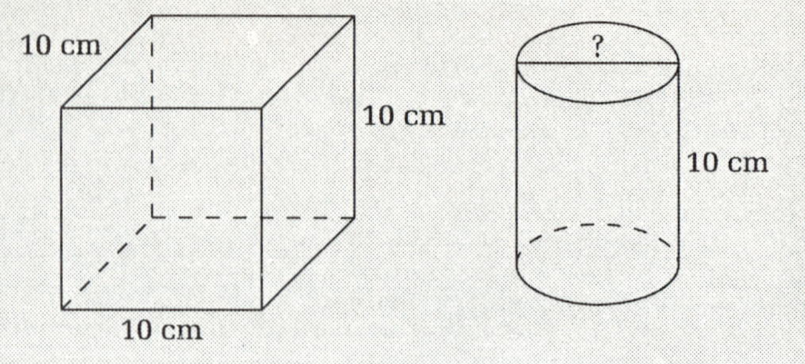

UNDERSTANDING THE MAIN IDEAS

Volume of a prism

The volume of a prism is found by multiplying the area of the base by the height of the prism.

Volume of a prism = area of base × height

$$V = Bh$$

Volume is measured in cubic units.

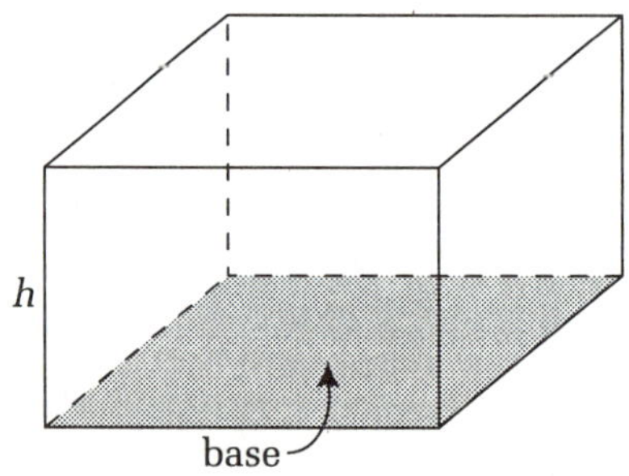

Sample 1

Find the volume of the triangular prism shown at the right.

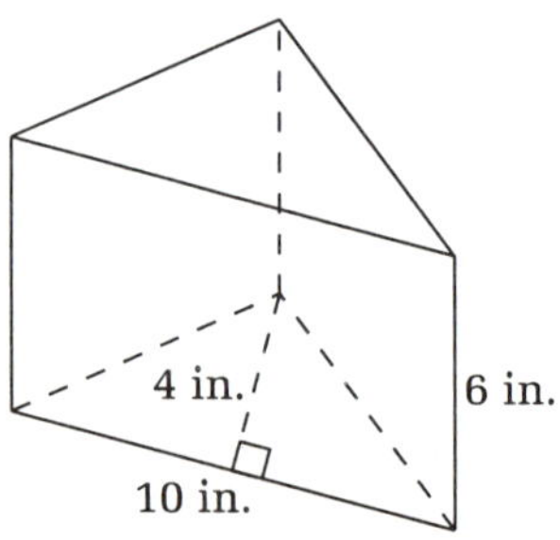

Sample Response

Use the formula $V = Bh$.

Step 1 Find the area of the base.

Since the base is a triangle, use the formula for the area of a triangle.

$A = \frac{1}{2}bh$

$= \frac{1}{2}(10)(4)$ ⟵ Substitute 10 for b and 4 for h.

$= 20 \text{ in.}^2$

The area of the base of the prism is 20 in.^{2}.

Sample Response continues on the next page.

Step 2 Find the volume of the prism.

$V = Bh$ — Substitute 20 for B and 6 for h. (*Caution:* In the volume formula, use the height of the prism, *not* the height of the triangular base.)

$= (20)(6)$

$= 120 \text{ in.}^3$

The volume of the prism is 120 in.^3.

Find the volume of each prism.

1.

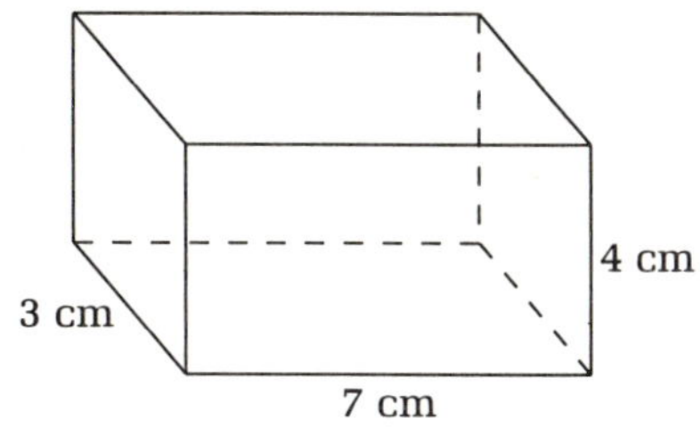

2.

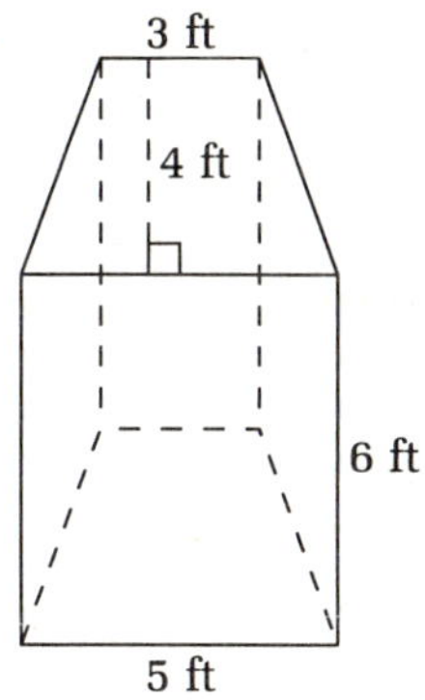

Sample 2

The volume of the trapezoidal prism shown at the right is 1944 ft^3. Find the height, h, of the prism.

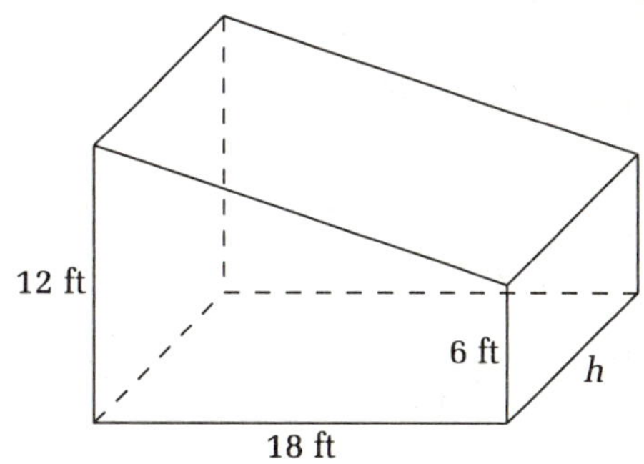

Sample Response

Use the formula $V = Bh$.

Step 1 Find the base area, B.

The dimensions of the trapezoidal bases are given. Use the formula for the area of a trapezoid.

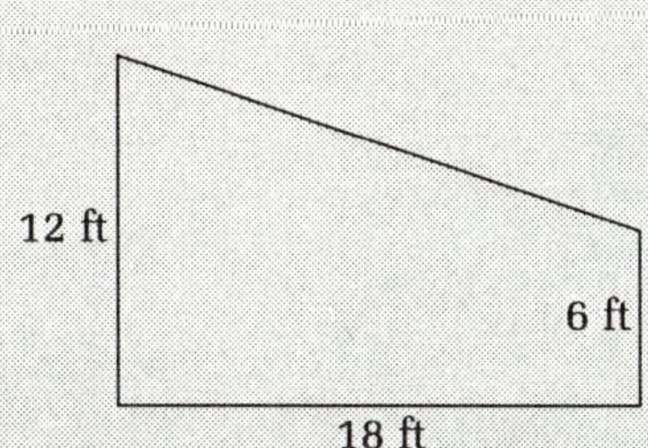

$A = \frac{1}{2}(b_1 + b_2)h$ ⟵ The variable h in this formula is the height of the base, *not* the height of the prism.

$= \frac{1}{2}(12 + 6)18$ ⟵ Substitute 12 for b_1, 6 for b_2, and 18 for h.

$= \frac{1}{2}(18)18$

$= 162 \text{ ft}^2$

Sample Response continues on the next page.

Step 2 Substitute the known values in the volume formula.

$$V = Bh$$
$$1944 = 162h \longleftarrow \text{Substitute 1944 for } V \text{ and 162 for } B.$$
$$\frac{1944}{162} = \frac{162h}{162}$$
$$12 = h$$

The height of the prism is 12 ft.

Find the missing dimension *h*.

3.

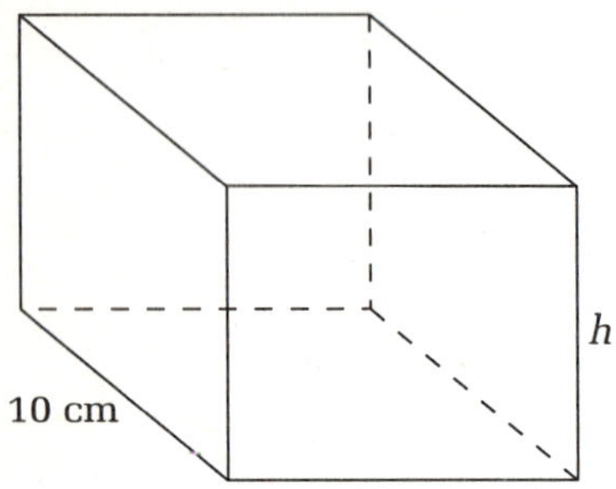

Volume $= 1430 \text{ cm}^3$

4.

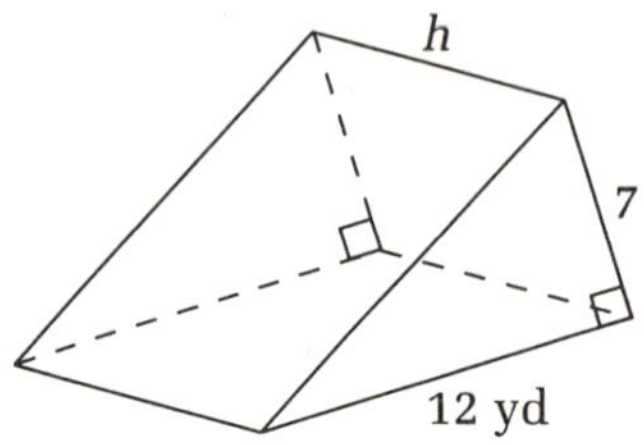

Volume $= 252 \text{ yd}^3$

Volume of a cylinder

The volume of a cylinder can also be found by using the formula $V = Bh$. For a cylinder, the value of B is the area of *one* of the circular bases, πr^2. As for a prism, the volume of a cylinder is measured in cubic units.

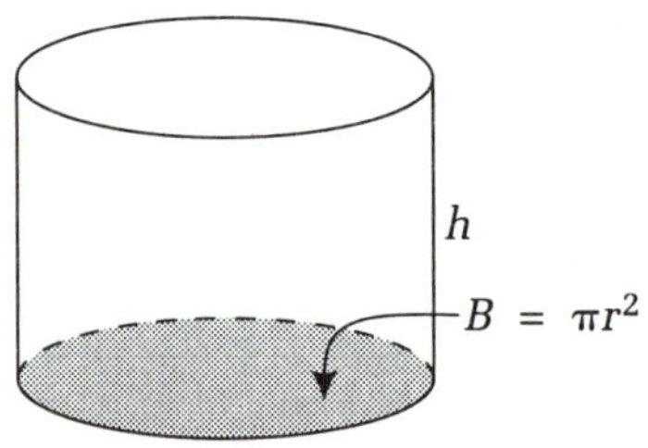

Sample 3

Find the volume of the cylinder shown at the right.

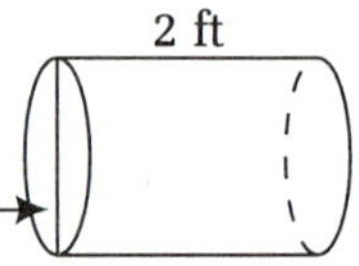

Sample Response

Use the formula $V = Bh$.

Step 1 Change all dimensions to inches: $h = 24$ in.

Step 2 Find the area of the base.

Use the formula for the area of a circle to find the area of a base. Since the diameter of the circular base is 18 in., the radius is $18 \div 2 = 9$ in.

$$A = \pi r^2$$
$$= \pi(9)^2 \longleftarrow \text{Substitute 9 for } r.$$
$$= \pi(81)$$
$$\approx 254.5 \text{ in.}^2$$

Sample Response continues on the next page.

Step 3 Find the volume of the cylinder.

$V = Bh$

$\approx 254.5(24)$ ⟵ Substitute 254.5 for B and 24 for h.

$\approx 6108.0 \text{ in.}^3$

The volume of the cylinder is about 6108.0 in.^3.

Find the volume of each cylinder.

5.

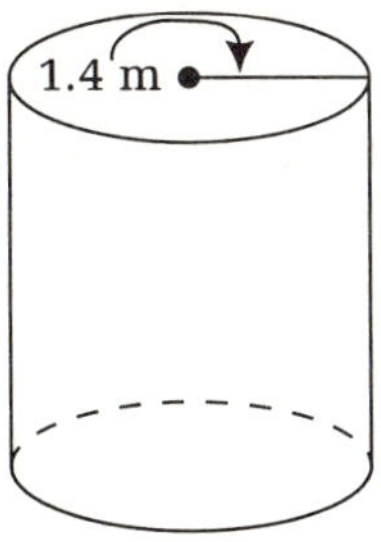

6.

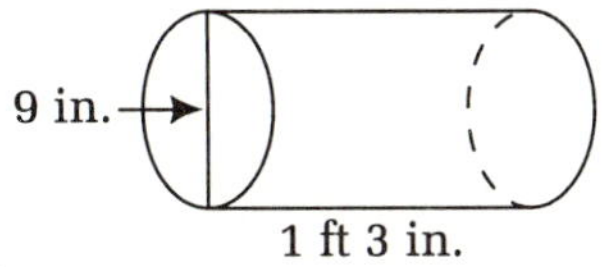

Find the missing dimension.

7.

Volume $\approx 1256.6 \text{ mm}^3$

8.

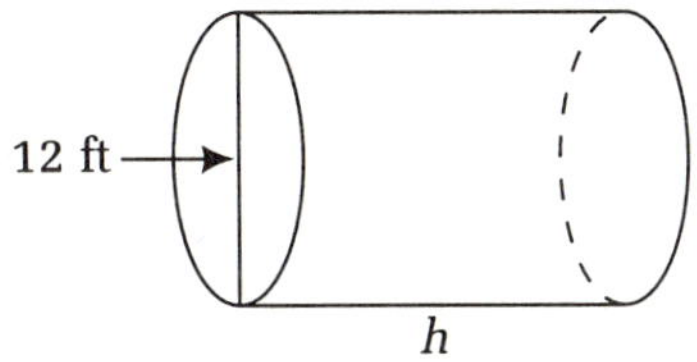

Volume $\approx 1809.6 \text{ ft}^3$

Review **PREVIEW**

For Exercises 9 and 10, identify each space figure and find the surface area.

9.

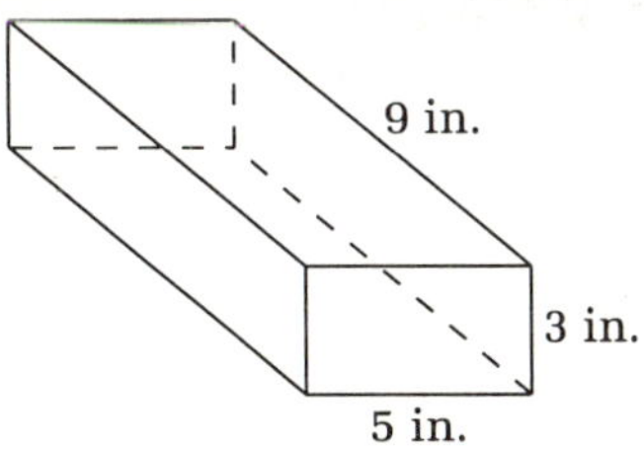

10.

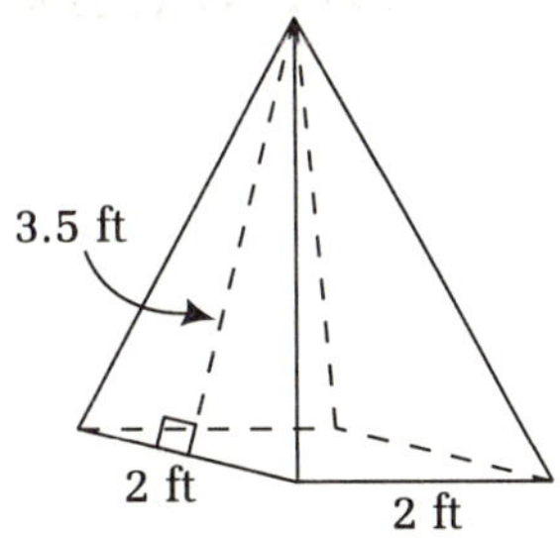

11. Convert 60 mi/h to feet per second.

12. The height and the length of a side of the base of the regular square pyramid at the right are labeled. Find the slant height, s, of the pyramid.

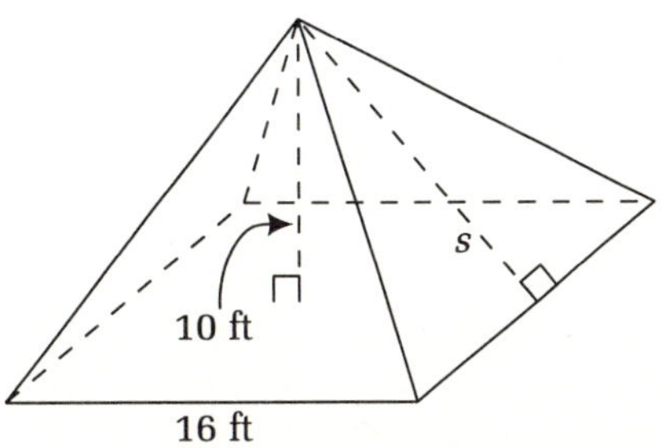

13. Two points on a line are (–3, 1) and (0, 7). Write an equation for the line.

Volumes of Pyramids and Cones

FOCUS

Find the volume of pyramids and cones.

When surf-fishing in the ocean from the beach, a fisherman uses special pyramid-shaped weights on his line to keep the waves from dragging his bait back to shore. He must choose a weight whose size and shape are appropriate for the surf conditions. The companies that manufacture these weights use the formula for the volume of a pyramid to determine the amount of metal needed to produce each weight.

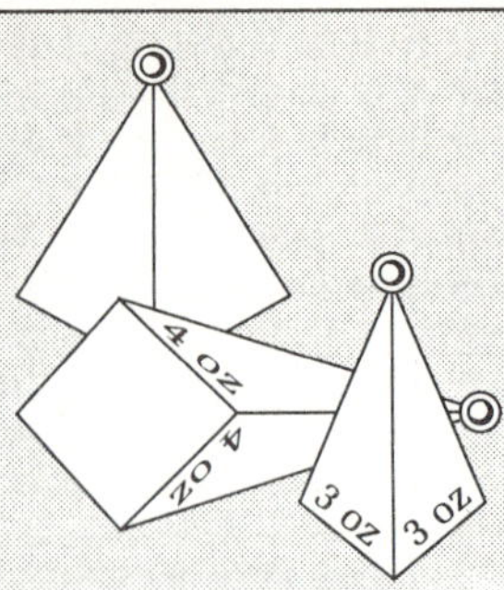

KEY TERMS	EXAMPLE / ILLUSTRATION
Cone a space figure with one circular base and a vertex	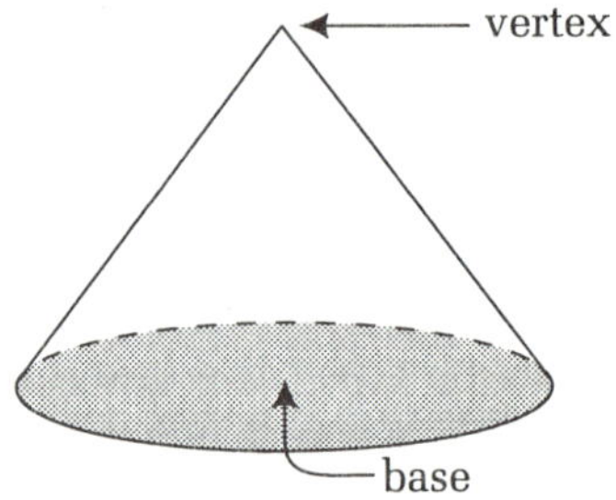

UNDERSTANDING THE MAIN IDEAS

Volume of a pyramid

The volume of a pyramid is $\frac{1}{3}$ the volume of a prism that has the same base area and the same height.

$$\text{Volume of a pyramid} = \frac{1}{3} \times \text{area of base} \times \text{height of the pyramid}$$

$$V = \frac{1}{3}Bh$$

The volume is measured in cubic units.

Sample 1

A pyramid has a square base 240 in. long on each side and a height of 160 in. Find the volume of the pyramid.

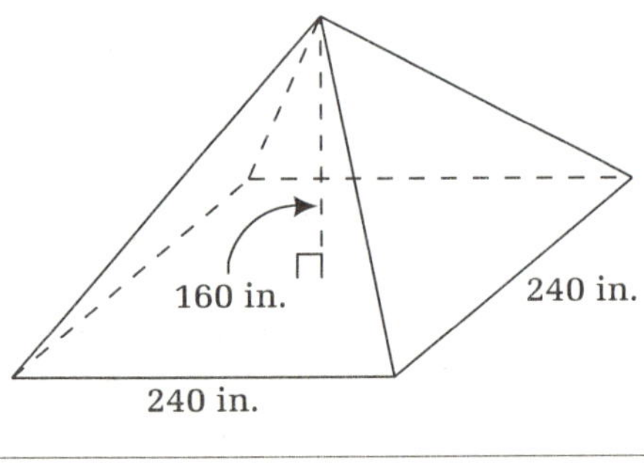

Sample Response

Use the formula $V = \frac{1}{3}Bh$.

Step 1 Find B, the area of the square base.
Use the formula for the area of a square.

$$\begin{aligned} A &= s^2 \\ &= (240)^2 \longleftarrow \text{Substitute 240 for } s. \\ &= 57{,}600 \text{ in.}^2 \end{aligned}$$

Step 2 Find the volume.

$$\begin{aligned} V &= \frac{1}{3}Bh \\ &= \frac{1}{3}(57{,}600)(160) \longleftarrow \text{Substitute 57,600 for } B \text{ and 160 for } h. \\ &= 3{,}072{,}000 \text{ in.}^3 \end{aligned}$$

The volume of the pyramid is 3,072,000 in.^{3}.

Find the volume of each pyramid.

1.

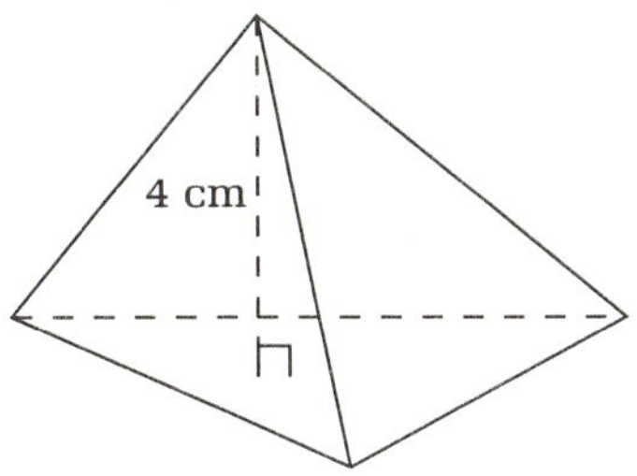

$B = 6.92 \text{ cm}^2$

2.

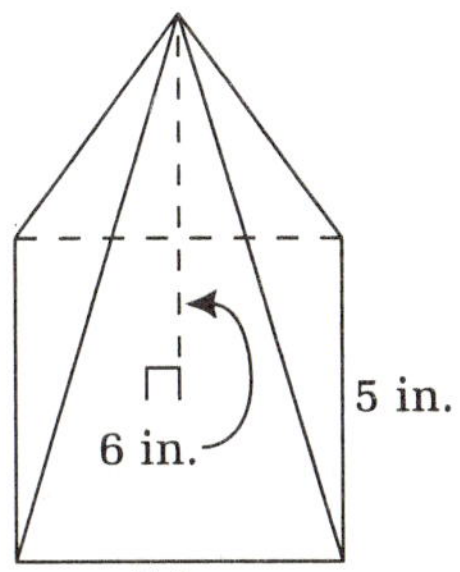

Volume of a cone

The volume of a cone is $\frac{1}{3}$ the volume of a cylinder that has the same base area and the same height.

$$\text{Volume of a cone} = \frac{1}{3} \times \text{area of base} \times \text{height of cone}$$

$$V = \frac{1}{3}Bh$$

The volume is measured in cubic units.

Sample 2

Find the volume of this cone-shaped drinking cup.

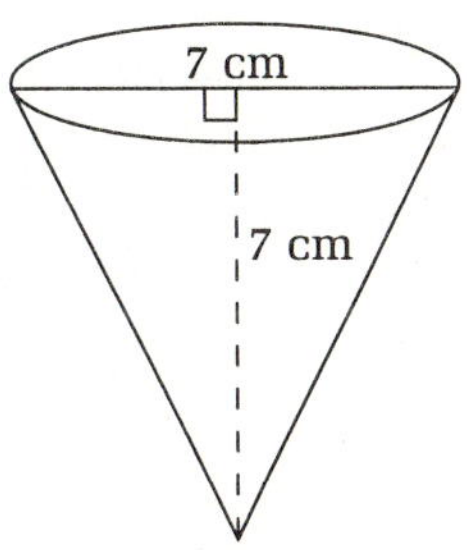

Sample Response

Use the formula for the volume of a cone.

Step 1 Find B, the area of the circular base.

Use the formula for the area of a circle. Since the diameter of the circular base is 7 cm, the radius is $7 \div 2 = 3.5$ cm.

$$\begin{aligned} A &= \pi r^2 \\ &= \pi(3.5)^2 \quad \longleftarrow \text{Substitute 3.5 for } r. \\ &\approx 38.5 \text{ cm}^2 \end{aligned}$$

Step 2 Find the volume.

$$\begin{aligned} V &= \frac{1}{3}Bh \\ &\approx \frac{1}{3}(38.5)7 \quad \longleftarrow \text{Substitute 38.5 for } B \text{ and 7 for } h. \\ &\approx 89.8 \text{ cm}^3 \end{aligned}$$

The volume of the drinking cup is about 89.8 cm^3.

Find the volume of each cone.

3.

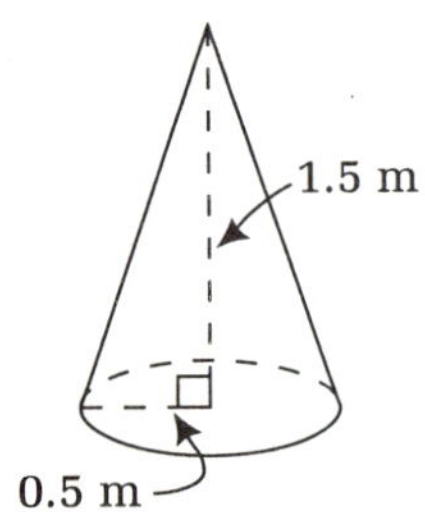

4.

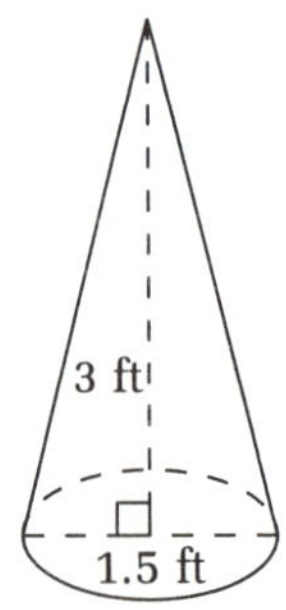

Review **PREVIEW**

Name the space figure and find the volume.

5.

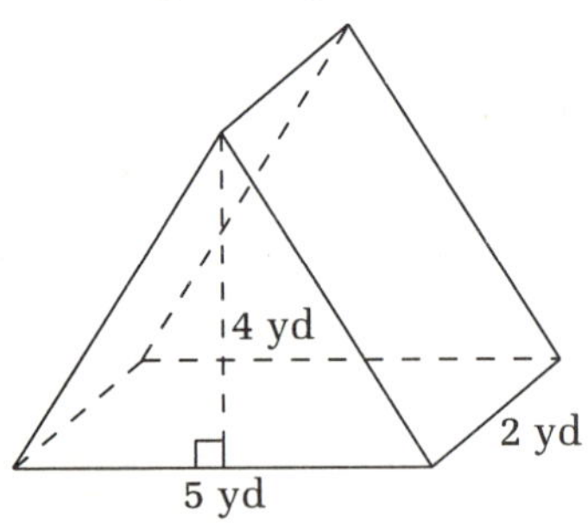

6.

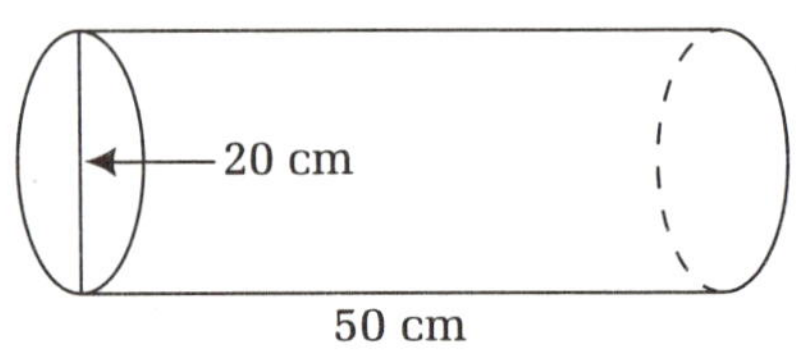

For Exercises 7–9, polygon *ABCDEF* ~ polygon *PQRSTU*.

7. What is the scale factor between the polygons?
8. Find BC.
9. Find ST.

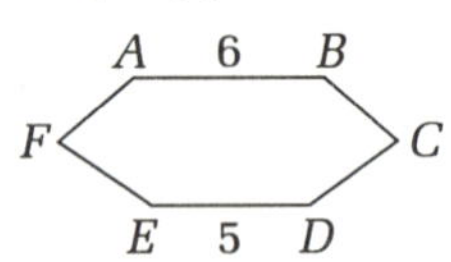

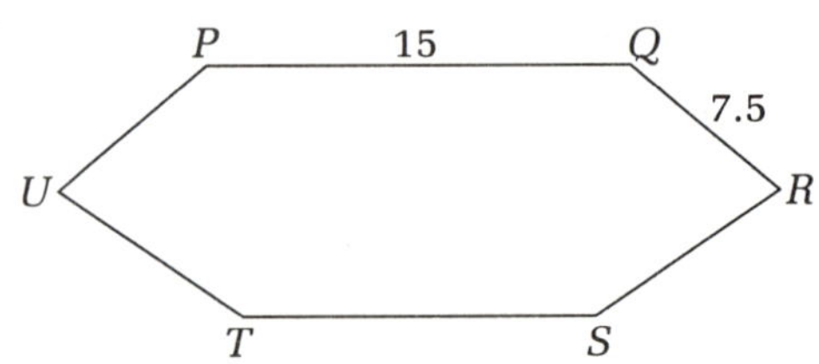

10. Solve the proportion $\frac{x^2}{9} = \frac{64}{81}$.

Sine and Cosine Ratios

FOCUS

Learn to use two special right triangle ratios, *sine* and *cosine*, to solve problem situations.

Trigonometric ratios are used in engineering, surveying, and navigation. For example, a pilot who wants to approach an airport at a 6° angle of descent can use trigonometry to determine when to start the descent.

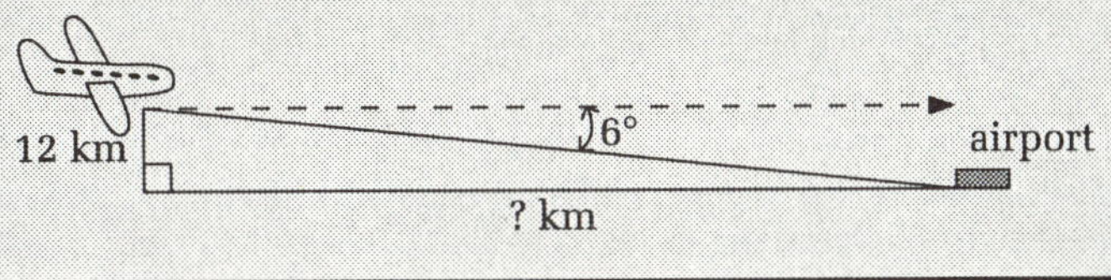

KEY TERMS	*EXAMPLE / ILLUSTRATION*
Sine the ratio of the length of the leg opposite an acute angle of a right triangle to the length of the hypotenuse (abbreviated *sin*)	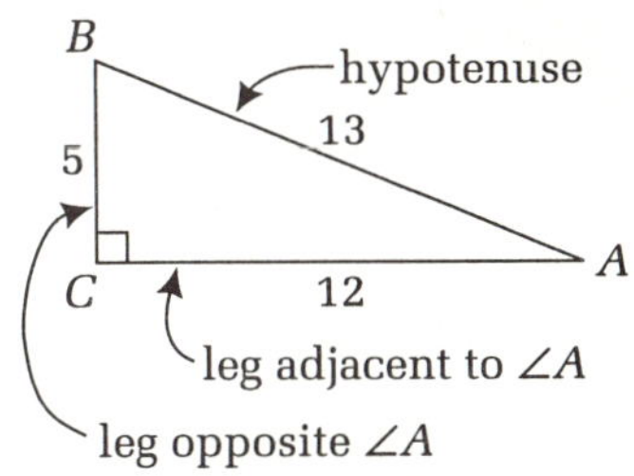
Cosine the ratio of the length of the leg adjacent to an acute angle of a right triangle to the length of the hypotenuse (abbreviated *cos*)	In $\triangle ABC$, $\sin A = \frac{5}{13}$ and $\cos A = \frac{12}{13}$. (There are no sine or cosine ratios associated with $\angle C$, because it is not acute.)
Trigonometric ratios ratios associated with the acute angles of a right triangle involving the lengths of sides	For $\triangle ABC$ above, $\sin A = \frac{5}{13}$, $\cos A = \frac{12}{13}$, $\sin B = \frac{12}{13}$, and $\cos B = \frac{5}{13}$.

UNDERSTANDING THE MAIN IDEAS

Sine and cosine ratios

The lengths of the legs and the length of the hypotenuse of a right triangle are used to find the sine and cosine of any acute angle.

Sample 1

Write each trigonometric ratio as a fraction and as a decimal rounded to hundredths.

a. $\sin Y$ **b.** $\cos Y$

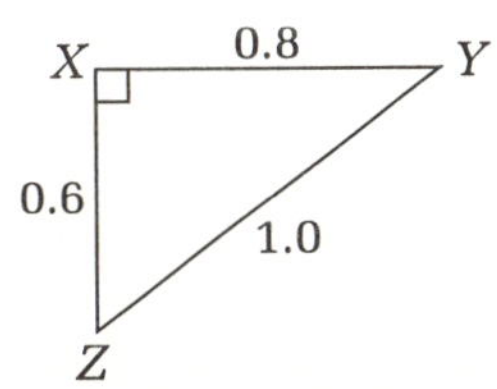

Sample Response

a. Use the definition of the sine ratio to find sin Y.

$$\sin Y = \frac{\text{length of leg opposite } \angle Y}{\text{length of hypotenuse}} = \frac{\text{length of } XZ}{\text{length of } YZ} = \frac{0.6}{1.0} = 0.6$$

b. Use the definition of the cosine ratio to find cos Y.

$$\cos Y = \frac{\text{length of leg adjacent to } \angle Y}{\text{length of hypotenuse}} = \frac{\text{length of } XY}{\text{length of } YZ} = \frac{0.8}{1.0} = 0.8$$

Use a calculator to write each ratio as a decimal rounded to hundredths. (Be sure your calculator is in degree mode.)

1. sin 36° **2.** cos 15° **3.** sin 15° **4.** cos 75°

For Exercises 5–10, use the figures at the right.

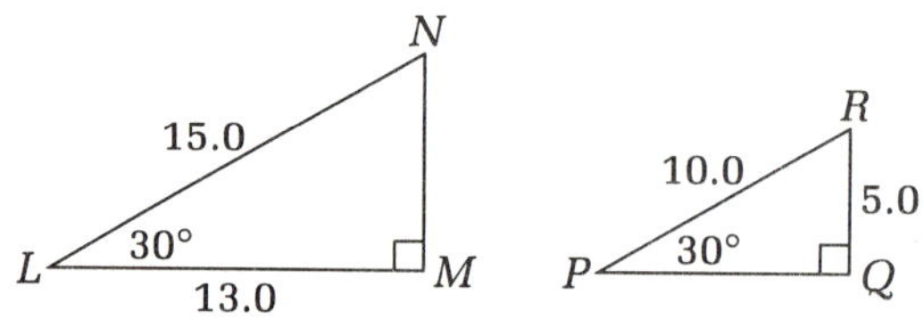

5. Why is $\triangle LMN \sim \triangle PQR$?

6. Use proportions to find MN and PQ to the nearest tenth.

7. Use ratios to find sin L and sin P.

8. Use ratios to find cos L and cos P.

9. Why are the answers in Exercise 7 and the answers in Exercise 8 the same?

10. Use a calculator to find sin 30° and cos 30°. Compare the results with your answers to Exercises 7 and 8.

Using trigonometric ratios to solve problems

You can use the sine and cosine ratios, which are found using calculators or tables, to solve problems involving right triangles.

Sample 2

An anchor is attached to a rowboat by a 20-yd rope. You drop the anchor and let the boat drift until the rope is taut. At that time, the rope makes a 28° angle with the surface of the water. To the nearest tenth of a yard, how deep is the lake at the point you dropped the anchor?

Sample Response

Make a sketch. Label the known lengths and angles. Let d = the depth of the water.

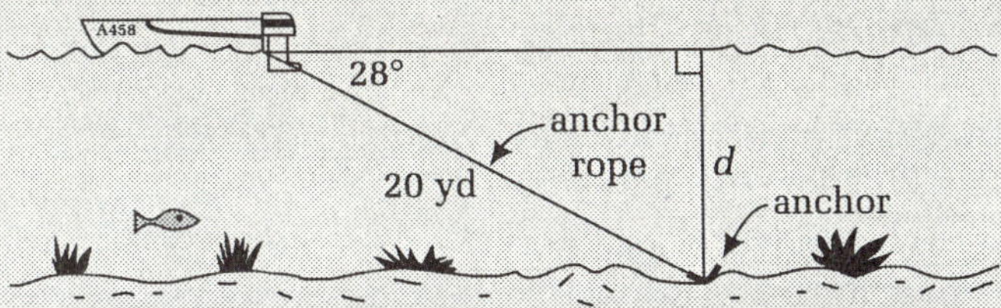

Notice that the known angle, 28°, the known length, 20 yd, and the unknown depth, d, are all parts of a right triangle. Write an equation involving a trigonometric ratio.

$\sin 28° = \frac{d}{20}$ ← Use sine because d is *opposite* the 28° angle.

Solve for d.

$20 \cdot \sin 28° = 20 \cdot \frac{d}{20}$

$20(0.4695) \approx d$ ← Use the SIN key on your calculator.

$9.39 \approx d$

The lake is about 9.4 yd deep where the anchor was dropped.

11. In Sample 2, how far has the boat drifted from the point where you dropped the anchor?

Write an equation you could use to find x. Then find x to the nearest tenth.

12.

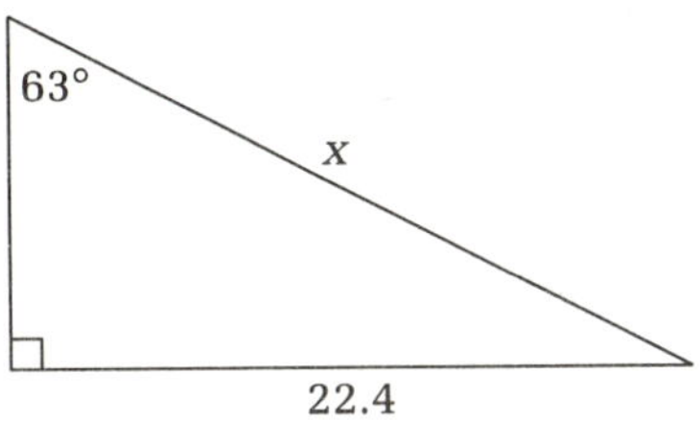

13.

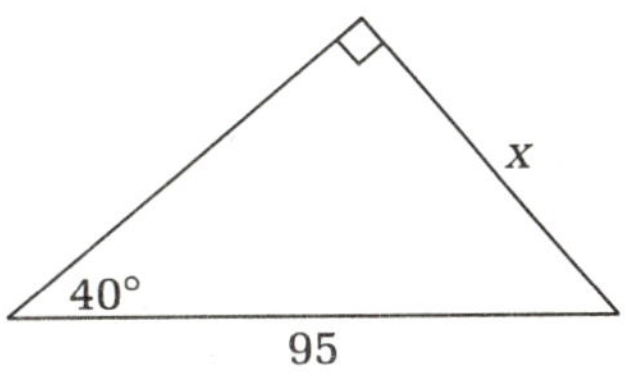

14. The diagonal of a rectangle makes a 37° angle with the base of the rectangle. The diagonal is 24 in. long. To the nearest tenth, what are the dimensions of the rectangle?

Review PREVIEW

15. Draw a dilation of $\triangle JKL$ (shown at the right) with center M and a scale factor of 2.

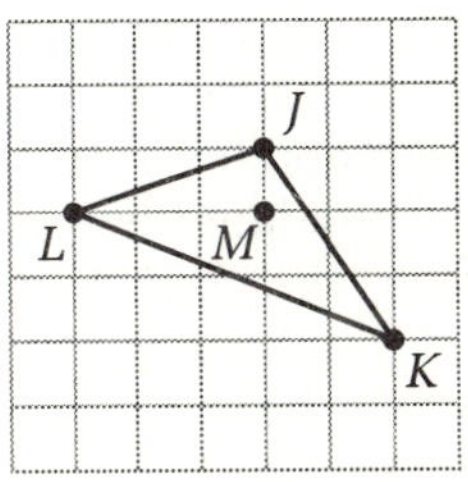

16. Describe one way that a graph can be misleading.

17. A box contains 8 blue marbles, 10 red marbles, and 6 green marbles. The marbles are all the same size. If a marble is picked at random, what is the probability that it is red?

Direct Variation, Slope, and Tangent

FOCUS

Explore three relationships that involve constant ratios.

In San Francisco, there are many steep streets, some with brick walls along them. You can judge the slope of such a street by counting bricks. (The bricks are $8\frac{1}{8}$ inches long and $2\frac{5}{8}$ inches wide, including the mortar.)

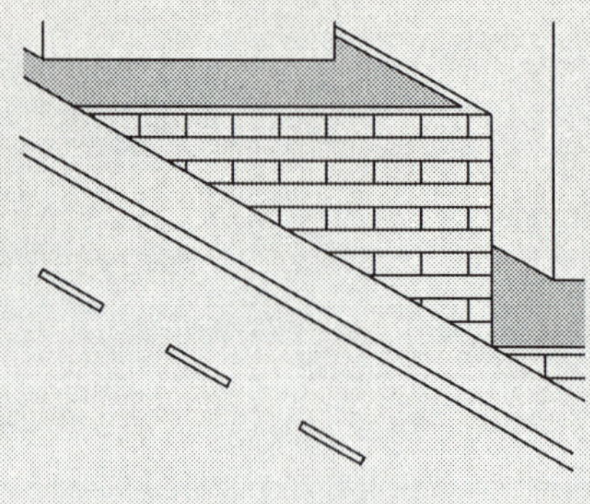

KEY TERMS	EXAMPLE / ILLUSTRATION
Direct variation the relationship between two variable quantities that have a constant ratio	<table><tr><th>a</th><th>b</th><th>c</th></tr><tr><td>2</td><td>0.6</td><td>0.3</td></tr><tr><td>3</td><td>0.9</td><td>0.3</td></tr><tr><td>5</td><td>1.5</td><td>0.3</td></tr></table>The quantities a and b are in direct variation because the ratio $\frac{b}{a}$ is always the same.
Variation constant the constant ratio in a direct variation	The variation constant in the example above is 0.3.
Slope of a line the ratio of rise to run for any two points on the line (The slope is a measure of the steepness of a line.)	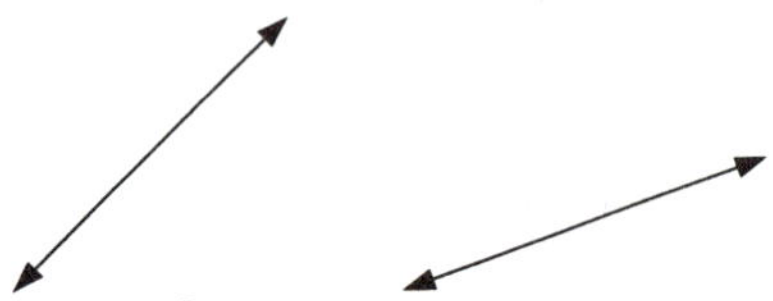
Rise the vertical change between two points **Run** the horizontal change between two points	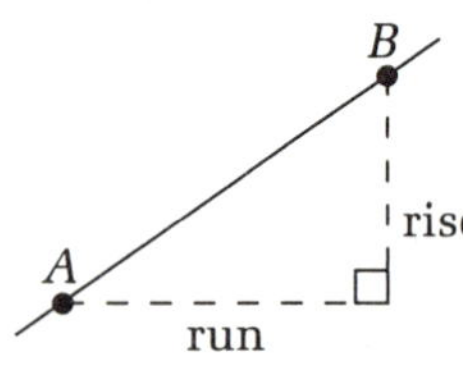
Tangent ratio a ratio associated with an acute angle of a right triangle that compares the length of the two legs of a right triangle (It is the ratio of the length of the leg opposite the angle to the length of the leg adjacent to the angle: $\tan A = \frac{\text{leg opposite } \angle A}{\text{leg adjacent to } \angle A}$.)	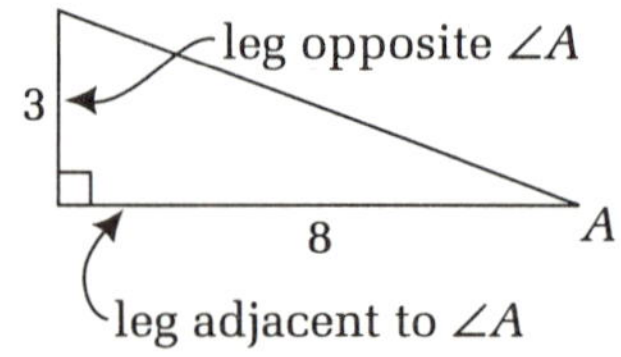$\tan A = \frac{3}{8}$

UNDERSTANDING THE MAIN IDEAS

Slope

Sample 1

Decide which of two straight highways is steeper: Route 1, which ascends 2500 ft in a horizontal distance of 10,000 ft, or Route 2, which climbs 2365 ft in a horizontal distance of 7500 ft.

Sample Response

Make a sketch.

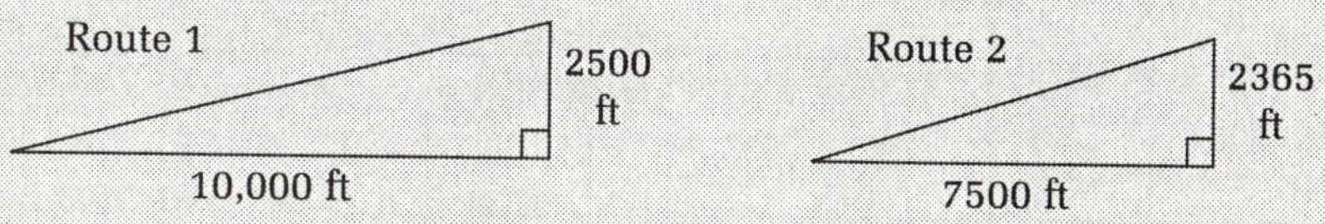

Calculate the slope of each route and compare the slopes.

$$\text{Slope of Route 1} = \frac{\text{rise}}{\text{run}} = \frac{2500}{10{,}000} = 0.25$$

$$\text{Slope of Route 2} = \frac{\text{rise}}{\text{run}} = \frac{2365}{7500} \approx 0.32$$

Since the slope of Route 2 is greater than the slope of Route 1, Route 2 is steeper. Notice that even though Route 1 climbs higher, it is not as steep as Route 2.

1. Suppose Route 1 continues horizontally for another 5000 ft. How much farther does the roadway rise?
2. The figure at the right shows the front view of a tent that is 4 ft high and 6 ft across. Find the slope of the sides of the tent. (*Hint:* The peak of the tent is centered over the width of the tent.)

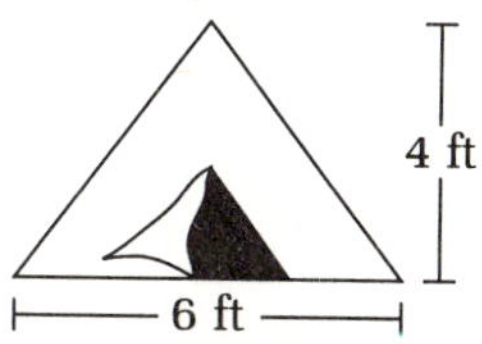

3. A ramp is to be built from the ground to a loading dock, which is 3 ft high. The slope of the ramp is to be 0.2. How far from the loading dock must the ramp begin?
4. The figure below shows a swimming pool that is 4 ft deep at the shallow end and 12 ft deep at the deep end. What is the slope of the floor of the pool?

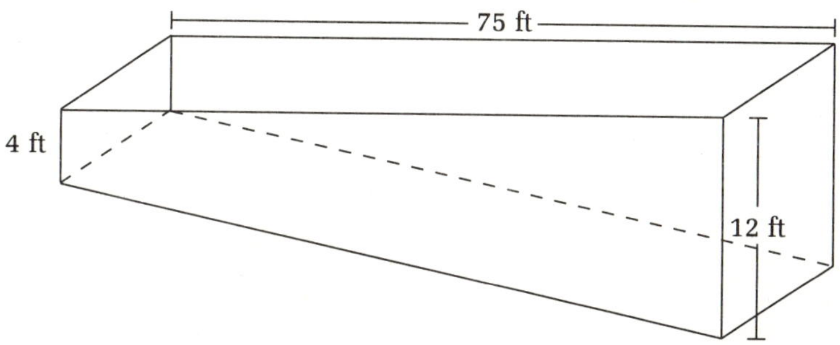

The tangent ratio

Sample 2

a. Find tan A to the nearest hundredth.

b. Use the inverse tangent key of a scientific calculator to find the measure of $\angle A$ to the nearest degree.

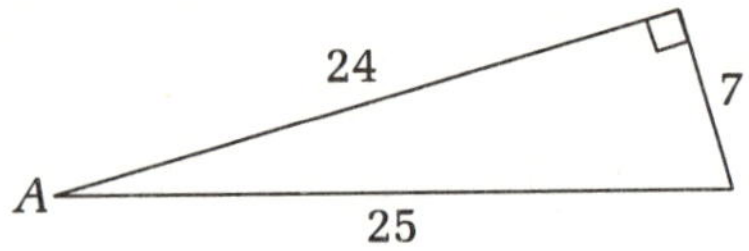

Sample Response

a. $\tan A = \dfrac{\text{leg opposite } \angle A}{\text{leg adjacent to } \angle A} = \dfrac{7}{24} \approx 0.29$

b. Press these keys on your calculator:

7 [÷] 24 [=] [INV] [TAN] ← *Caution:* Be sure the calculator is in *degree* mode.

The result, 16.2602..., means that $A = 16°$ to the nearest degree.

For Exercises 5–7,

a. find tan A to the nearest hundredth.

b. find the measure of $\angle A$ to the nearest degree.

5.

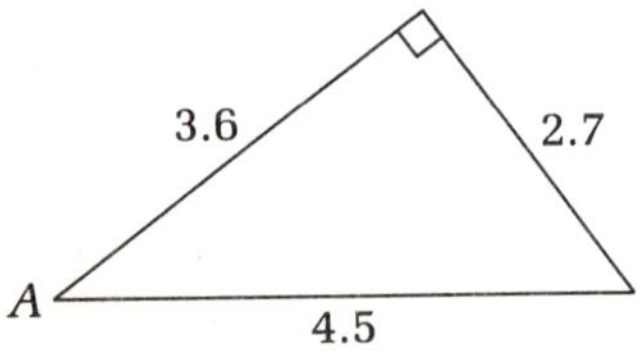

6.

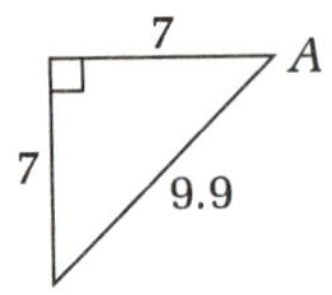

7.

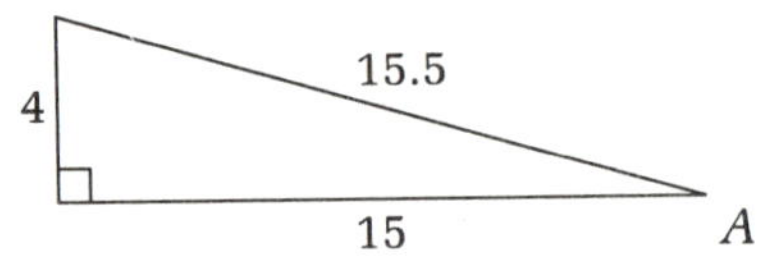

Find the length of $\overline{XY}$ to the nearest tenth.

8.

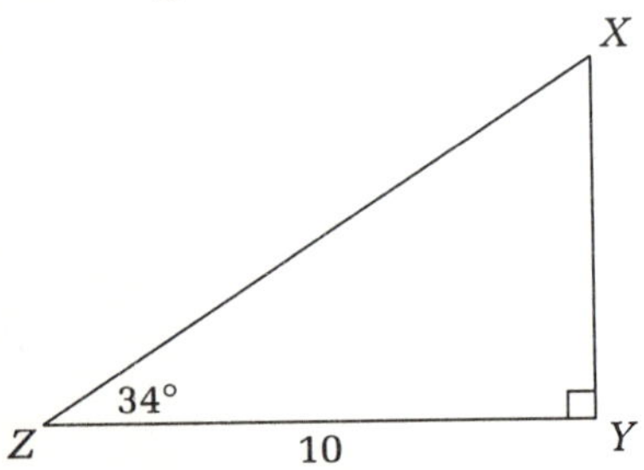

9.

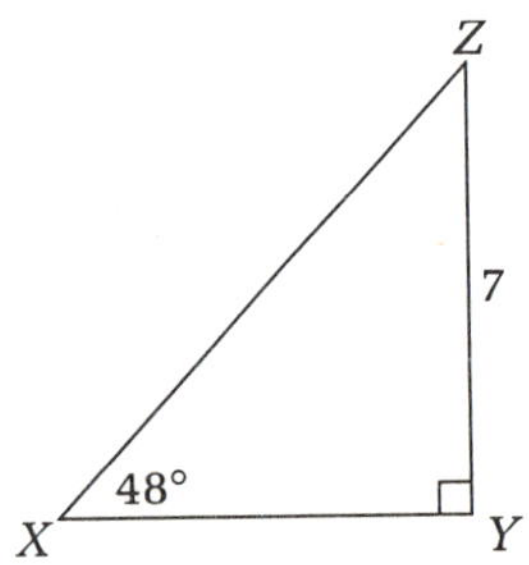

Review PREVIEW

10. Use $\triangle QRS$ to find sin Q and cos Q.

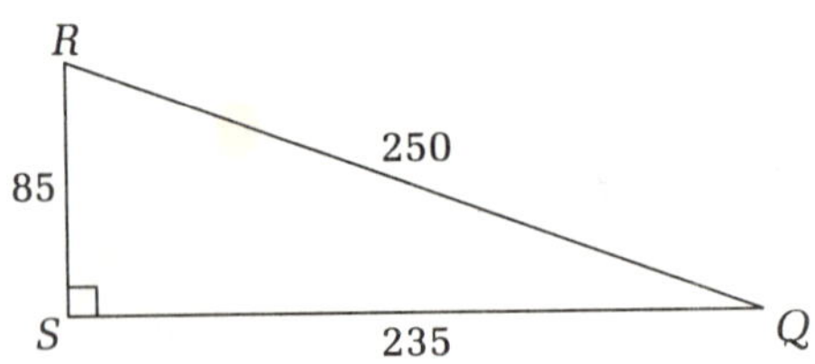

Solve.

11. $5x = 13$

12. $\dfrac{x}{13} = 5$

13. $5 = 13x$

14. Write an equation to model this statement:
A child's height at age 2 is about half the height the child will reach as an adult.

Translations

FOCUS

Translate figures on a coordinate plane, recognize translational symmetry.

Kathy Homstad of the United States takes her luge down a straightaway at 60 miles per hour. At this part in the race, she is not turning, only her position is changing. The luge is sliding on the ice. This is an example of a ***translation****.*

KEY TERMS	EXAMPLE / ILLUSTRATION
Translation the movement of an object without changing its size or shape and without rotating or flipping it	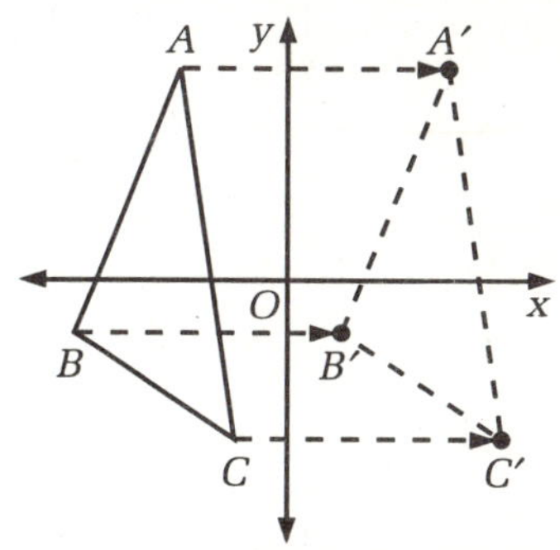
Translational symmetry a symmetry that occurs when an object or pattern is translated, creating copies of the object or pattern at various locations	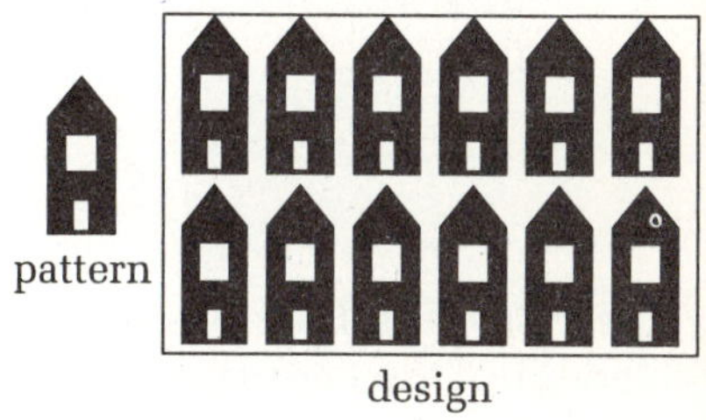

UNDERSTANDING THE MAIN IDEAS

Working with translations

Sample 1

Translate $\triangle ABC$ 5 units right and 4 units up. What are the coordinates of each vertex after the translation?

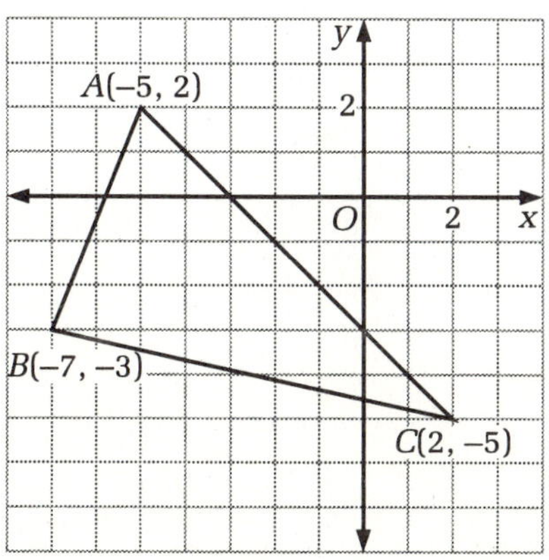

Sample Response

Translate each vertex 5 units right and 4 units up. Then draw the triangle.

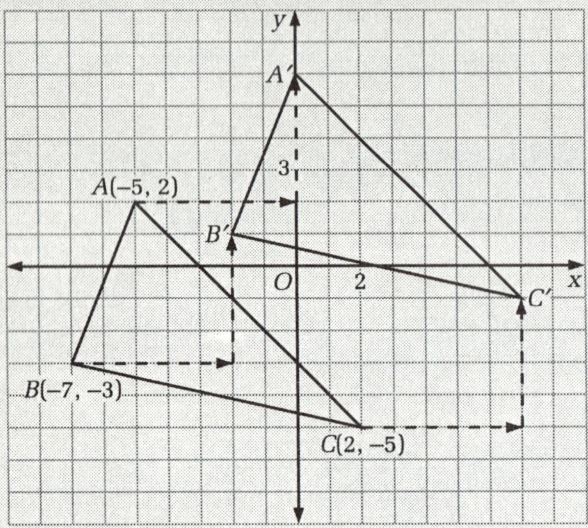

The translation above is written this way: $\triangle ABC \rightarrow \triangle A'B'C'$. The arrow means "goes to." The translation of vertices can be represented in this fashion: $A(-5, 2) \rightarrow A'(0, 6)$; $B(-7, -3) \rightarrow B'(-2, 1)$; $C(2, -5) \rightarrow C'(7, -1)$.

The coordinates of the vertices after the translation are $A'(0, 6)$, $B'(-2, 1)$, and $C'(7, -1)$.

For Exercises 1 and 2, translate $\triangle DEF$ as directed. Find the coordinates of each vertex after each translation.

1. 2 units right and 1 unit up
2. 4 units left and 5 units down

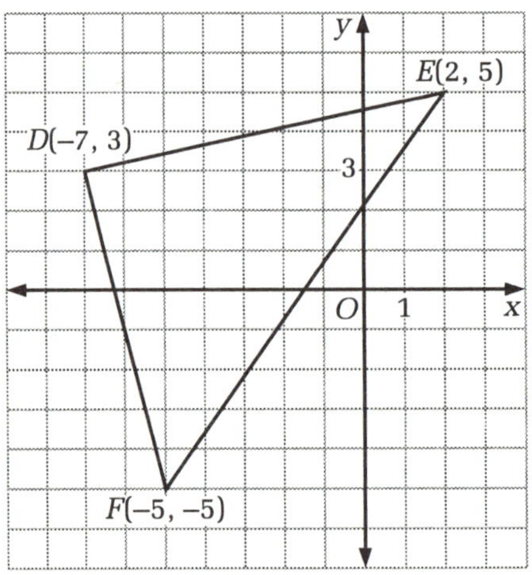

Tell whether the picture shows a translation. If so, describe how the vertices moved.

3.

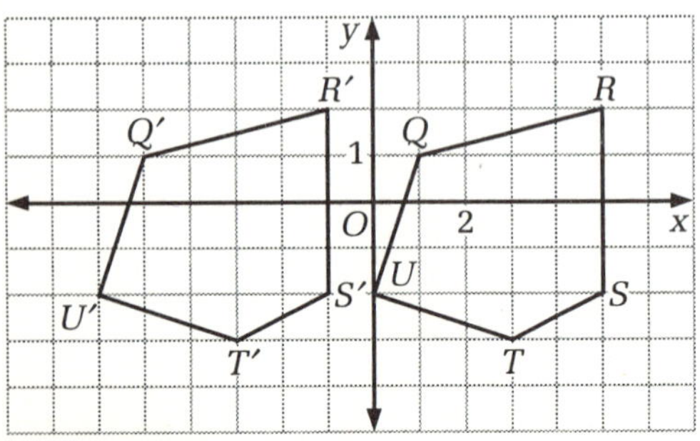

4.

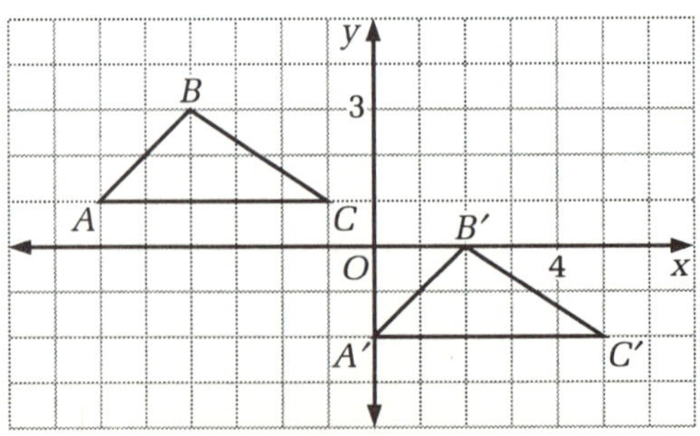

Describing translations using (x, y) coordinates

Another way to describe a translation is to show the change in the coordinates (x, y) of any point.

Sample 2

Describe the following translation by showing the change in the coordinates (x, y) of any point of a figure:

3 units right and 4 units down.

Sample Response

The words *right* and *up* are represented by addition and the words *left* and *down* are represented by subtraction. Since the movements "right" and "left" happen in the direction of the x-axis, the x-coordinate is changed by these movements. Similarly, the y-coordinate is changed by the movements "up" and "down."

Therefore, the given translation indicates an addition to the x-coordinate and subtraction from the y-coordinate. The translation can be written: $(x, y) \rightarrow (x + 3, y - 4)$.

Describe each translation by showing the change in the coordinates (x, y) of any point.

5. 6 units left and 4 units up

6. 2 units right and 7 units down

Sample 3

Write the coordinates of P' after the translation $(x, y) \rightarrow (x - 5, y + 4)$ of the point $P(2, 3)$.

Sample Response

In the translation, substitute 2 for x and 3 for y and then solve.

$(x - 5, y + 4) \rightarrow (2 - 5, 3 + 4) = (-3, 7)$

Write the coordinates of P′ after each translation of P(–3, 5).

7. $(x, y) \rightarrow (x + 5, y - 4)$

8. $(x, y) \rightarrow (x - 2, y)$

9. $(x, y) \rightarrow (x - 10, y + 15)$

10. $(x, y) \rightarrow (x, y + 6)$

Review **PREVIEW**

11. a. Plot the points $A(-1, -1)$, $B(4, 4)$, $C(0, 8)$, and $D(-4, 4)$. Connect them in order and connect D to A.

b. Write the specific name of the polygon you formed. Explain how you know what it is.

Solve each equation.

12. $x + 12 = 19$

13. $a - 5 = -15$

Rotations

FOCUS

Rotations of figures on a coordinate plane, rotate figures on polar graph paper, recognize rotational symmetry.

The pirouette is an example of a rotation which occurs in ballet. Notice the center of rotation.

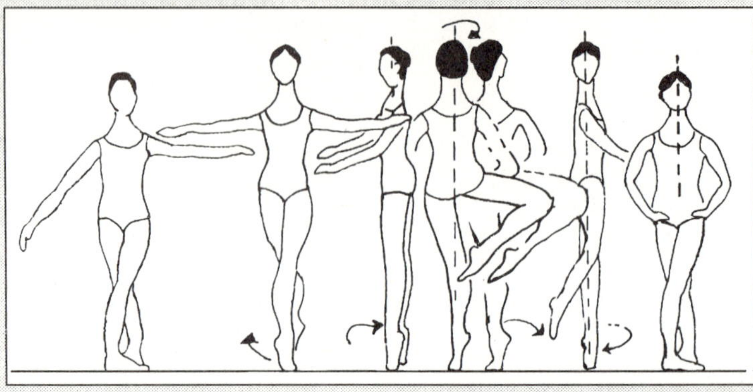

KEY TERMS	EXAMPLE / ILLUSTRATION
Rotation a turn around a point called the center of rotation (The direction of a rotation may be clockwise or counterclockwise.)	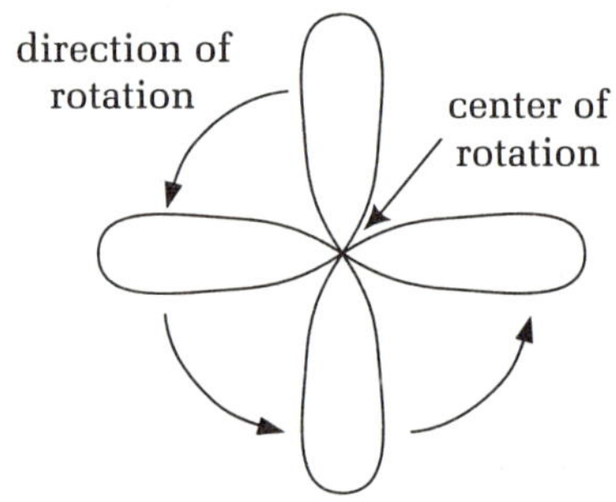
Transformation a change made to an object or its position (Translating and rotating are two ways to transform an object.)	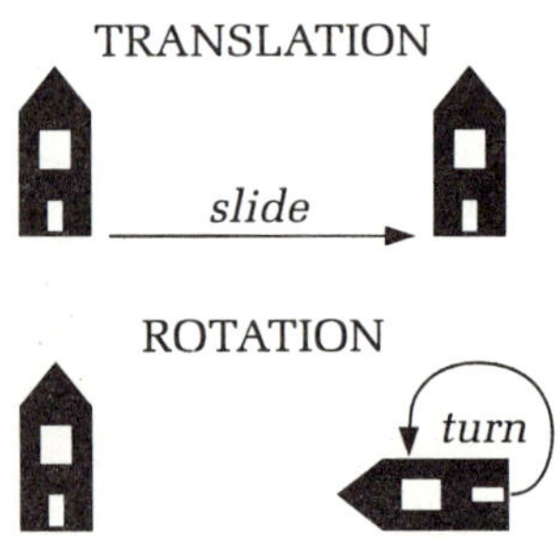
Rotational symmetry the symmetry that occurs when an object looks the same after a rotation *of less* than 360° (All objects look the same after they have been rotated 360°—this does not indicate rotational symmetry.)	An equilateral triangle has 120° and 240° rotational symmetry. 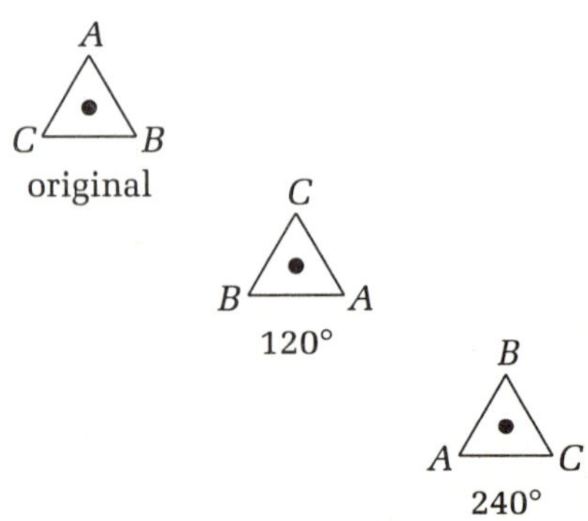

UNDERSTANDING THE MAIN IDEAS

Describing rotation

Sample 1

The graphs below show two different rotations of rectangle *ABCD* around the origin. Describe the direction and amount of each rotation.

a.

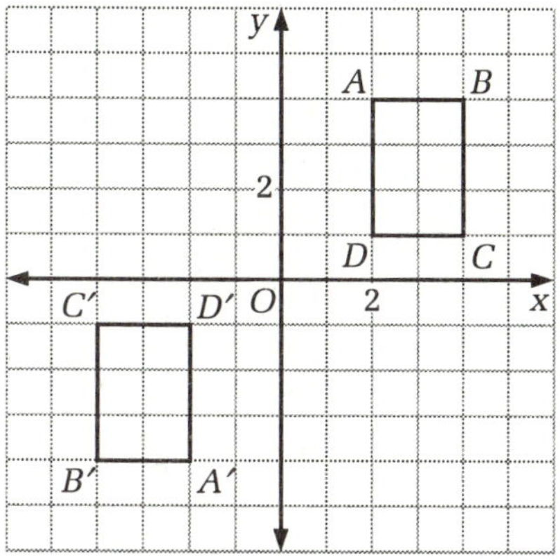

b.

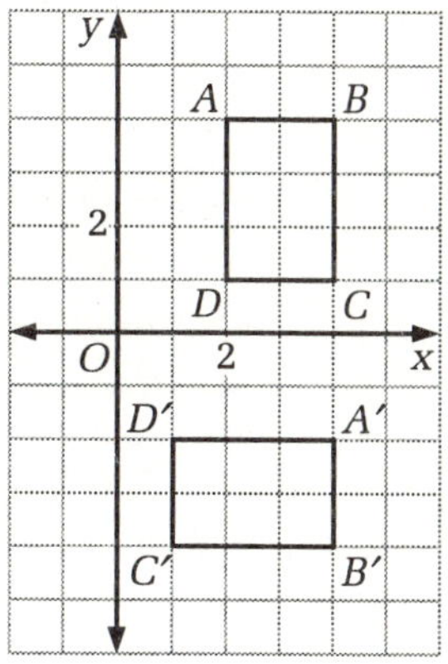

Sample Response

For each graph, draw a line from the center of rotation (in this case the origin) to a point on the original figure. Then draw a line from the center of rotation to the corresponding point on the rotated figure. Measure the angle formed. To completely describe a rotation, you need to state the amount of turn (as a fraction of a full turn or in degrees), the direction (clockwise or counterclockwise), and the center of rotation.

a.

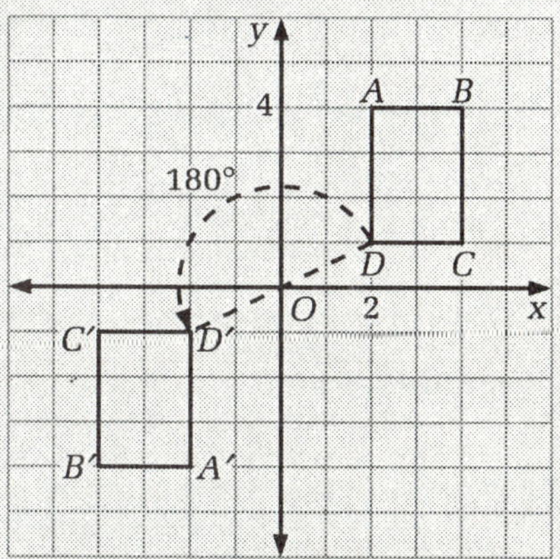

$180°$ or $\frac{1}{2}$ turn around the origin (Direction does not matter.)

b.

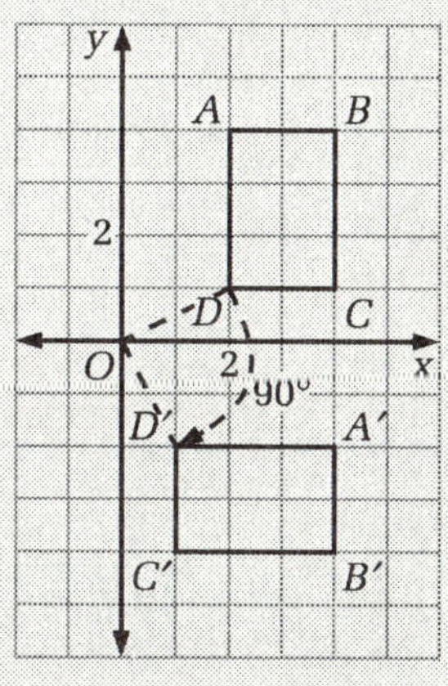

$90°$ or $\frac{1}{4}$ turn clockwise around the origin

The rotation in part (b) could be described as 270° counterclockwise. As a rule, rotations are described by the smaller of the two possible angles.

Graph the given figure and its image on a coordinate plane. Then describe the direction and amount of the rotation of each graph.

1. $\triangle ABC$: $A(3, -2)$, $B(1, -6)$, $C(8, -8)$; $\triangle A'B'C'$: $A'(2, 3)$, $B'(6, 1)$, $C'(8, 8)$
2. trapezoid *DEFG*: $D(-9, -2)$, $E(-2, -2)$, $F(-3, -6)$, $G(-6, -6)$;
   trapezoid *D′E′F′G′*: $D'(-2, 9)$, $E'(-2, 2)$, $F'(-6, 3)$, $G'(-6, 6)$

Drawing rotations

It is easier to draw rotations on a special type of graph paper called polar graph paper.

Sample 2

Rotate $\triangle QRS$ 80° clockwise around the origin.

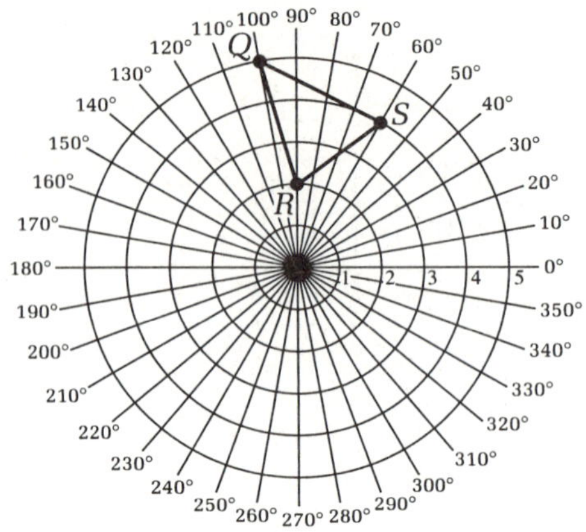

Sample Response

Move each vertex along its circle in a clockwise direction. The amount of turn is 80° which is 8 × 10°, so move the vertices eight lines. Draw the rotated triangle by connecting the rotated vertices.

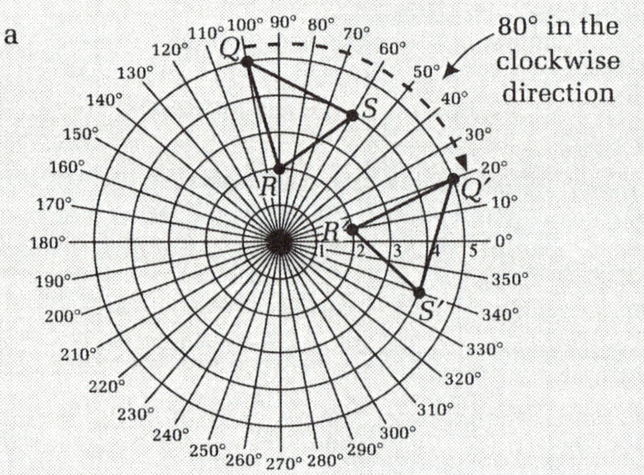

Copy each figure on polar graph paper. Draw each indicated rotation of the figure around the origin.

3. 150° counterclockwise

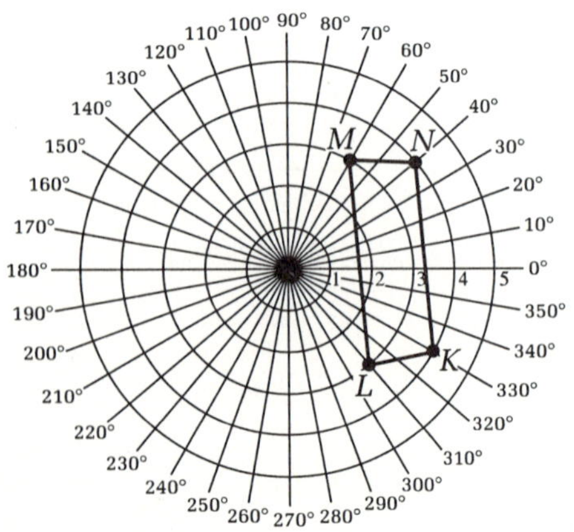

4. 90° clockwise

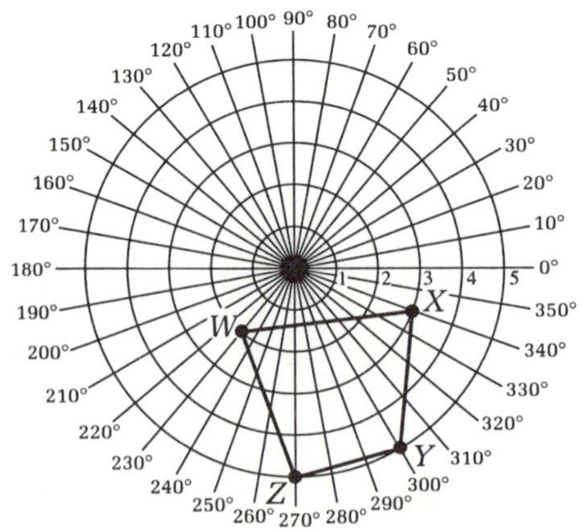

Sample 3

Tell whether or not each figure has rotational symmetry. If it does, describe the symmetry.

a. X

b. 

Sample Response

a. Yes. This letter has 180° rotational symmetry.

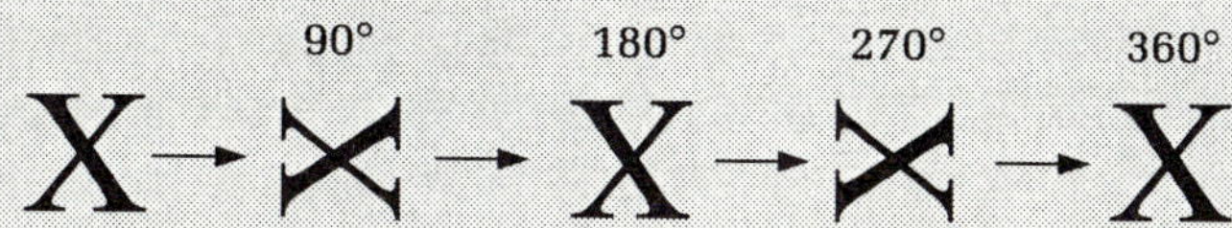

b. Yes. This design has 120° and 240° rotational symmetry.

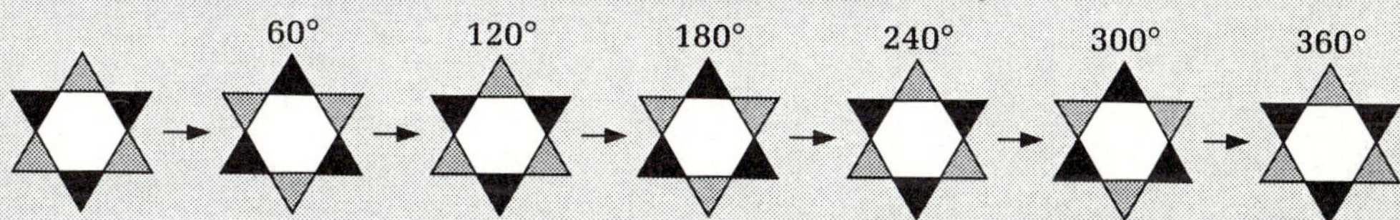

Tell whether each figure has rotational symmetry. If it does, describe the symmetry.

5. 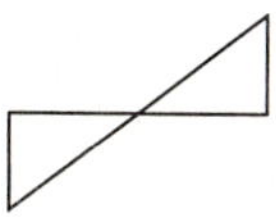

6. 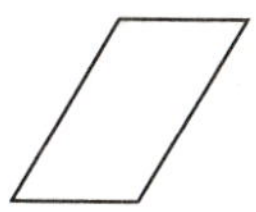

7. 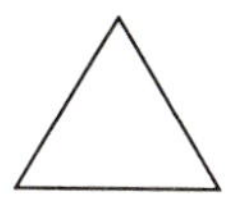

Review PREVIEW

8. Translate $\overline{AB}$ 5 units to the right and 4 units down. Draw $\overline{A'B'}$. What are the coordinates of each point after the translation?

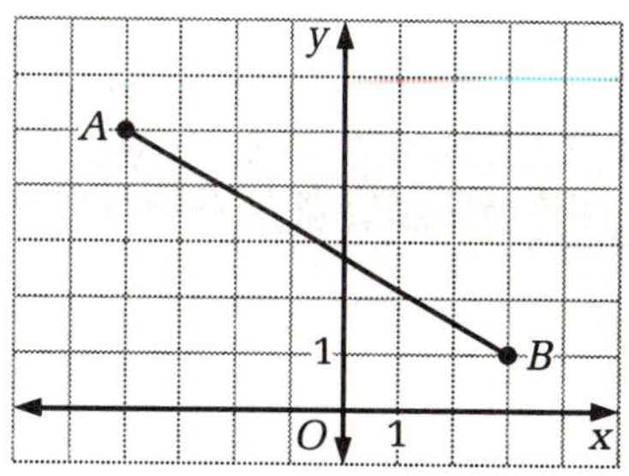

9. Make a graph to show the data.

National League Baseball Team ERA (July 10, 1993)

Atlanta	Chicago	Cincinnati	Colorado	Florida	Houston	Los Angeles
3.19	4.14	4.15	5.83	3.84	3.83	3.55
Montreal	**New York**	**Philadelphia**	**Pittsburgh**	**San Diego**	**San Francisco**	**St. Louis**
3.87	4.43	3.83	4.82	4.18	3.38	3.67

Dilations

FOCUS

Change the size of a figure without changing its shape.

Inflating the balloon causes the picture to be enlarged while keeping the same shape. This transformation is a dilation.

KEY TERMS	EXAMPLE / ILLUSTRATION
Image the result of a transformation	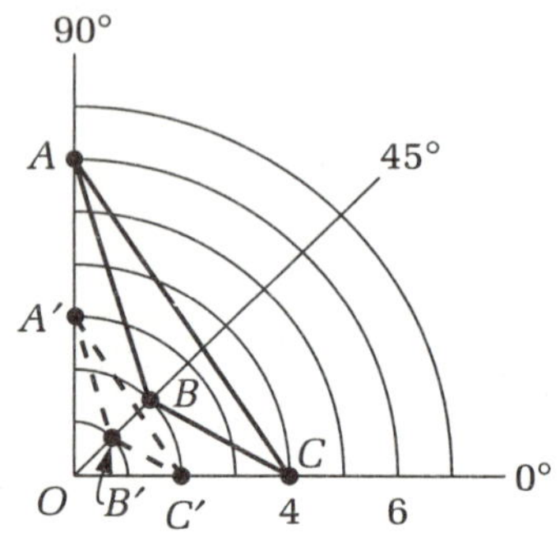 In the figure, the smaller triangle, $\triangle A'B'C'$, is the *image* of the larger triangle, $\triangle ABC$.
Dilation a transformation in which the original figure and its image are similar (When the image is larger than the original, the dilation is called an *enlargement*. When the image is smaller, the dilation is called a *reduction*.)	The transformation above is a *dilation*. $\triangle A'B'C' \sim \triangle ABC$. Since the image is smaller than the original, the dilation is a *reduction*.
Scale factor in a dilation, the ratio of any length on the image to the corresponding length on the original	The scale factor of the dilation above is $\frac{1}{2}$ since each side of the image is half as long as the corresponding side in the original.
Center of dilation the point which does not move under a dilation	The center of dilation of the transformation above is the origin O. Notice that $OA' = \frac{1}{2}OA$, $OB' = \frac{1}{2}OB$, and $OC' = \frac{1}{2}OC$.

UNDERSTANDING THE MAIN IDEAS

In a dilation, each image point lies on the ray from the center of dilation through the original point. In an enlargement, the image points are farther from the center than the originals. In a reduction, the image points are closer to the center.

Sample

Copy quadrilateral *ABCD* on graph paper and find its image under the dilation with center $O(0, 0)$ and scale factor 3.

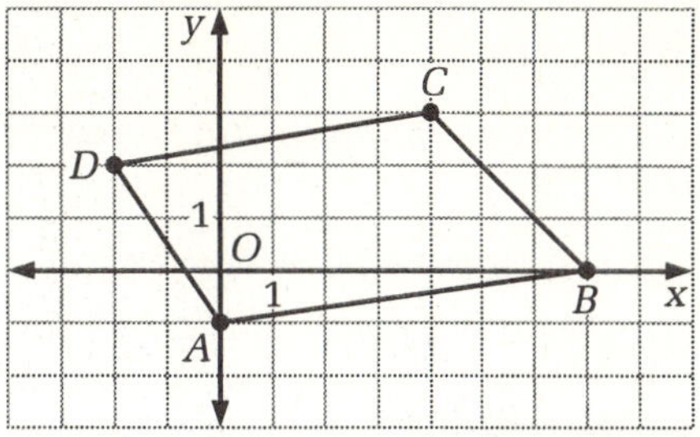

Sample Response

After copying quadrilateral *ABCD*, use a ruler to draw rays from the center through each vertex. On each ray, place the image point of the vertex so that it is three times as far from the center as the original point.

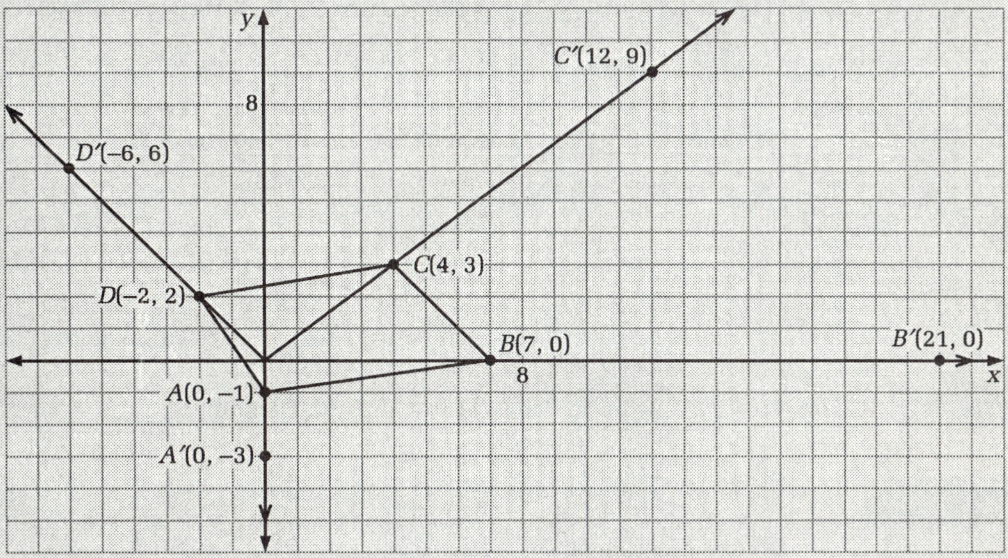

Finally, connect the image points to form the image quadrilateral.

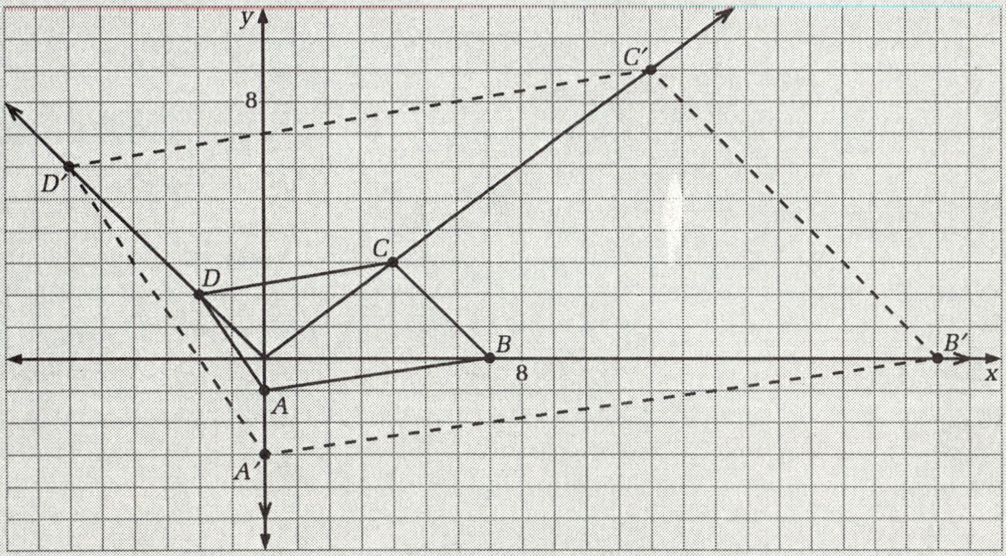

1. Use a protractor to measure all the angles in quadrilaterals *ABCD* and *A′B′C′D′* in the Sample. Are the corresponding angles congruent?

2. Copy quadrilateral *ABCD* on another sheet of graph paper. Draw the dilation of *ABCD* with center at (0, –2) and scale factor 3.

3. Are the images in the Sample and in Exercise 2 similar? Are they congruent?

4. On graph paper, draw $\triangle JKL$ with vertices *J*(0, 6), *K*(4, 6) and *L*(6, 0). Draw the dilation of $\triangle JKL$ with center (–2, 0) and scale factor $\frac{1}{3}$.

5. Draw $\triangle JKL$ from Exercise 4 again. Draw the dilation of $\triangle JKL$ with center *K* and scale factor $\frac{1}{2}$.

6. Are the images in Exercises 4 and 5 similar? Are they congruent?

In each figure, $\triangle A'B'C'$ is the image of $\triangle ABC$. Give the coordinates of the center and the scale factor.

7.

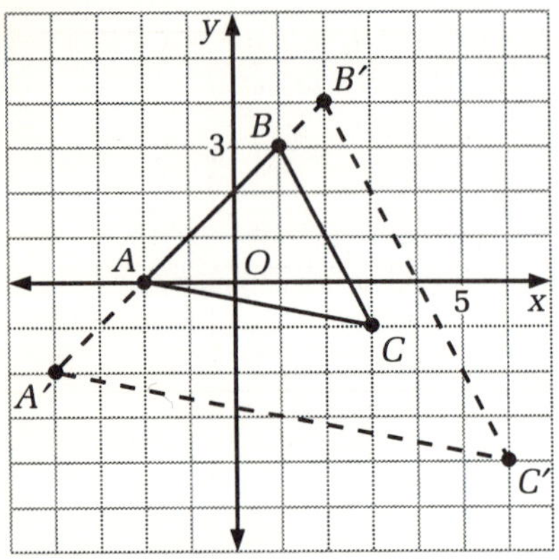

8.

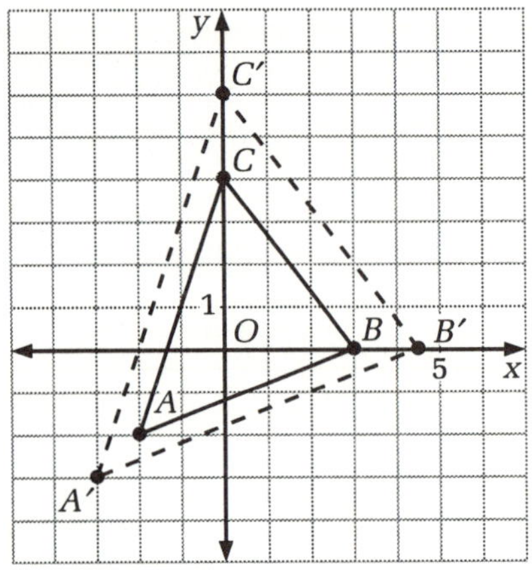

Review **PREVIEW**

9. Find all the missing side and angle measures. In the figure, $\triangle RST \sim \triangle XYZ$.

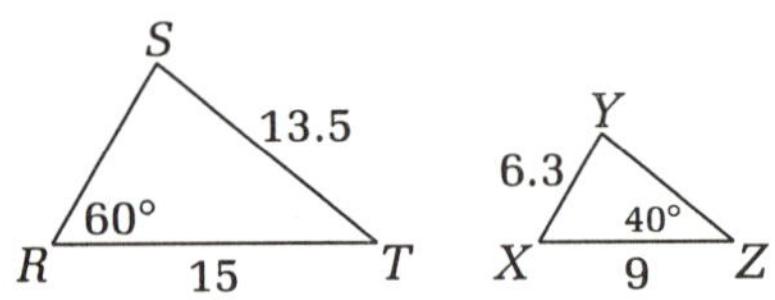

Solve.

10. $5x + 6 = x - 10$

11. $2n + 13 = 5n - 23$

12. $2a - 11 = -3a + 4$

13. $4w + 11 = -2w - 13$

Write each ratio in simplest form.

14. $\frac{12}{60}$

15. 18 to 72

16. 26 : 91

Reflections

FOCUS

Recognize and draw the mirror image of an object.

At this point in the game of billiards shown at the right, the purpose of the next shot is to hit the cue ball C in such a way that it makes no contact with ball A but hits ball B, sending ball B in the direction of the pocket in the lower right corner. The diagram shows how to use a cushion shot in which the cue ball C hits the side of the table and bounces back to hit ball B. The idea of reflection in a line helps you find the point along the side of the table at which you must aim. The reflection of ball B outside the table, with the edge of the cushion as the line of reflection, is used to locate this point. When cue ball C hits the cushion at the proper point, it rebounds off the cushion at an angle toward ball B.

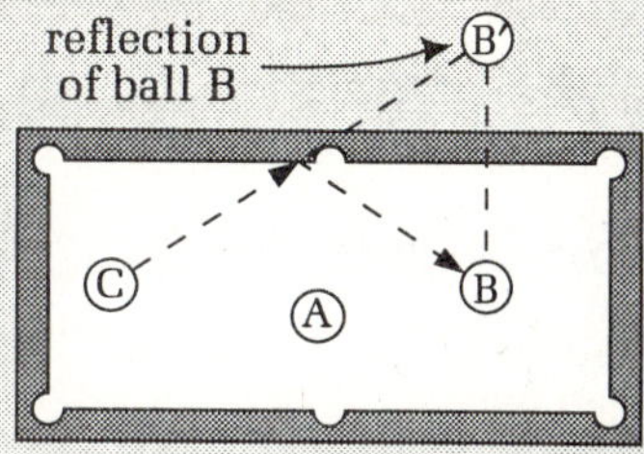

KEY TERMS	EXAMPLE / ILLUSTRATION
Reflection a transformation that produces a mirror image of a figure	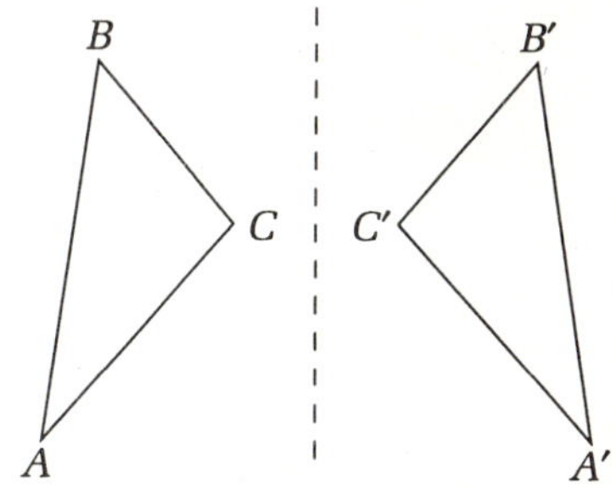 $\triangle A'B'C'$ is a reflection of $\triangle ABC$.
Line of reflection the line over which a figure is reflected	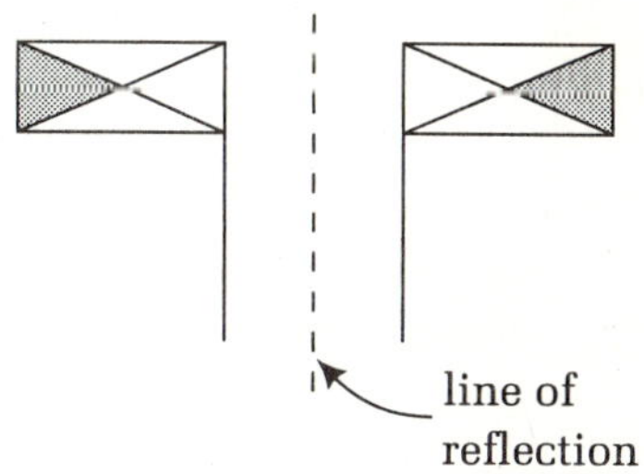
Orientation the direction, either clockwise or counterclockwise, in which the points on a figure are ordered (In a reflection, the orientation is reversed.)	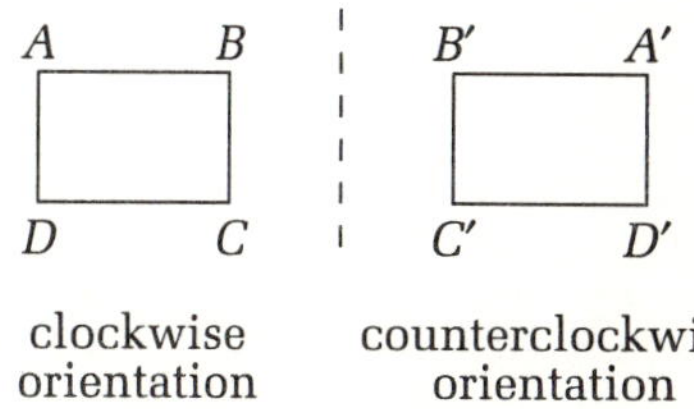 clockwise orientation; counterclockwise orientation

x-axis
the horizontal axis in the coordinate plane

y-axis
the vertical axis in the coordinate plane

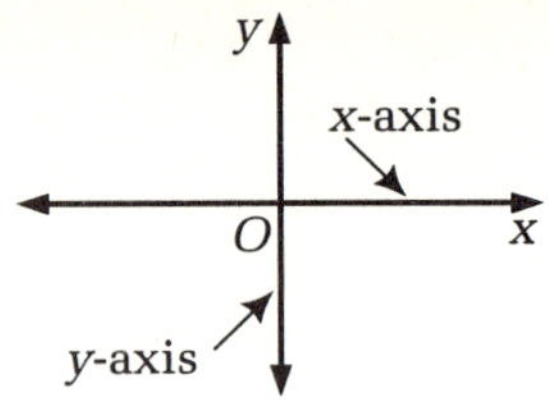

UNDERSTANDING THE MAIN IDEAS

Reflections

A reflection of a figure is a mirror image of the figure. The figure and its image are congruent ($\triangle ABC \cong \triangle A'B'C'$) and are located on opposite sides of the line of reflection. Each point on the image (such as point B') is the same distance from the line of reflection as the corresponding point (B) on the original figure. The orientation of the image (A'–B'–C' is counterclockwise) is in the opposite direction from the orientation of the original figure (A–B–C is clockwise).

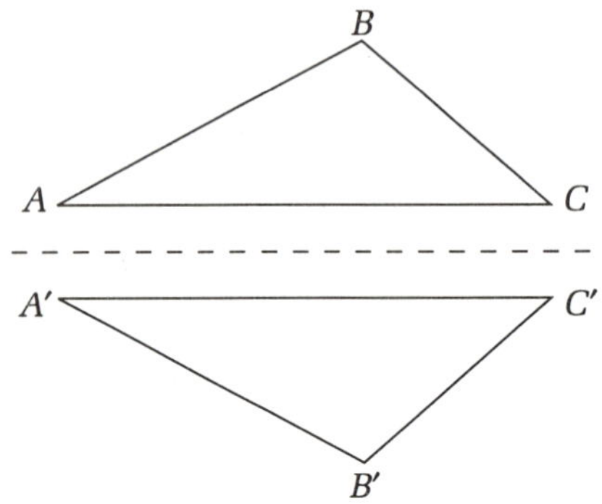

Sample 1

$\triangle A'B'C'$ is the reflection image of $\triangle ABC$. Copy the triangles, draw the line of reflection, and name three segments that are divided in half by the line of reflection. Describe the orientation of each triangle.

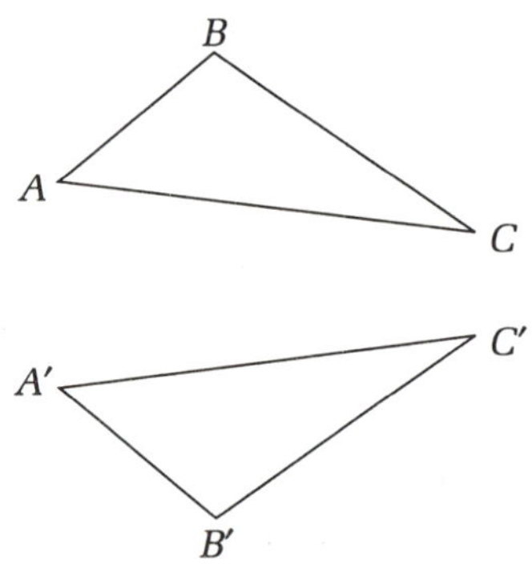

Sample Response

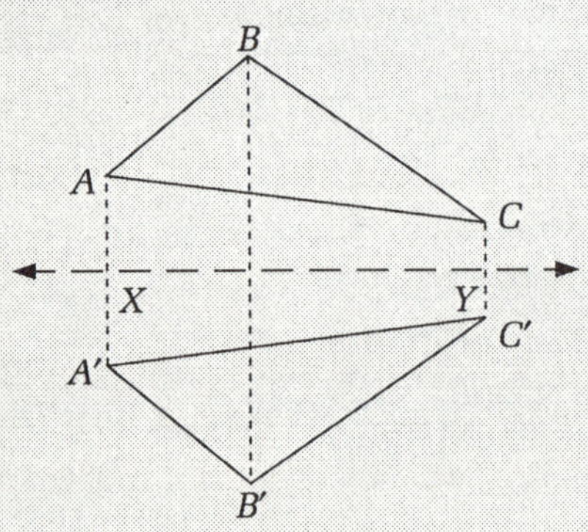

$\overleftrightarrow{XY}$ is the line of reflection; $\overline{AA'}$, $\overline{BB'}$, and $\overline{CC'}$ are divided in half by $\overleftrightarrow{XY}$.

$\triangle ABC$ has clockwise orientation;
$\triangle A'B'C'$ has counterclockwise orientation.

In Exercises 1–4, tell whether each transformation is a reflection. If it is not, tell why not.

1.

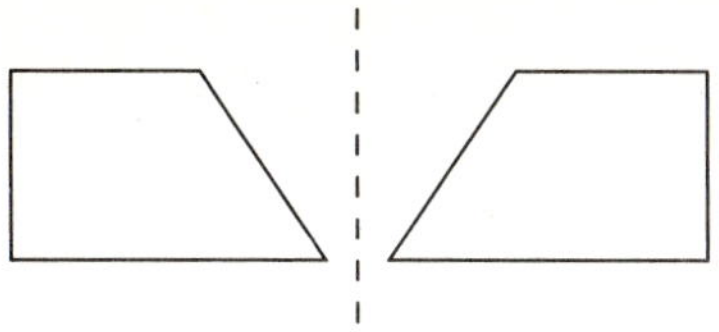

2.

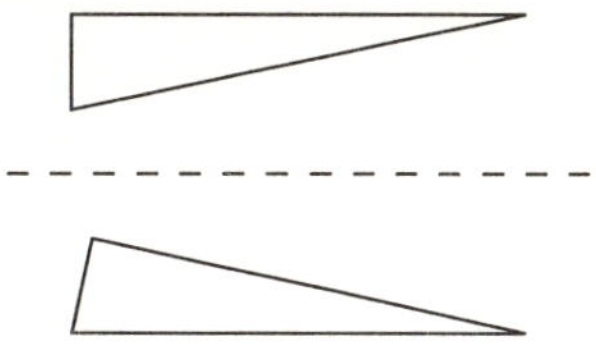

3.

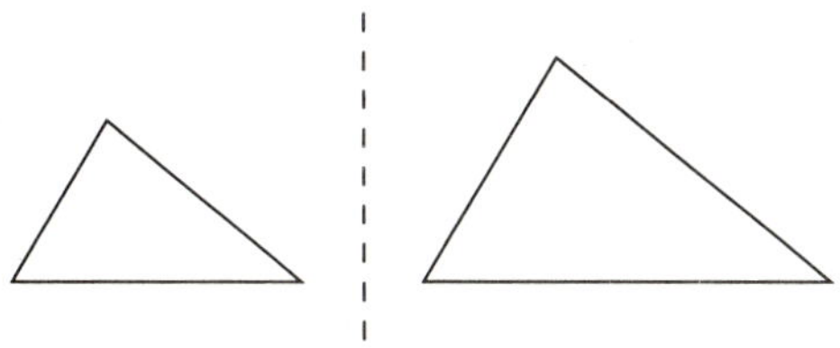

4.

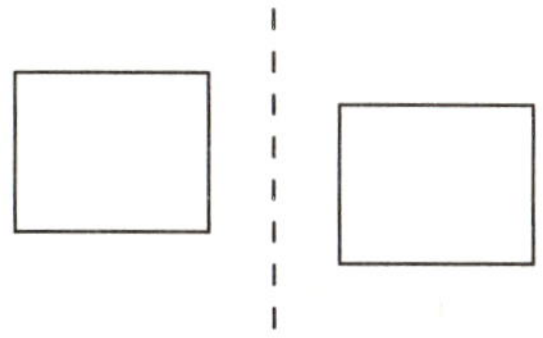

Reflections on a coordinate plane

When a polygon on a coordinate plane is reflected over the x-axis or the y-axis, the coordinates of the vertices of the polygon are useful when drawing the image of the original polygon.

Sample 2

Copy quadrilateral $WXYZ$ on graph paper. Indicate the x- and y-axes and give the coordinates of the vertices. Then reflect quadrilateral $WXYZ$ over the y-axis. Write the coordinates of the vertices of quadrilateral $W'X'Y'Z'$ and describe the orientation of quadrilateral $WXYZ$ and its image.

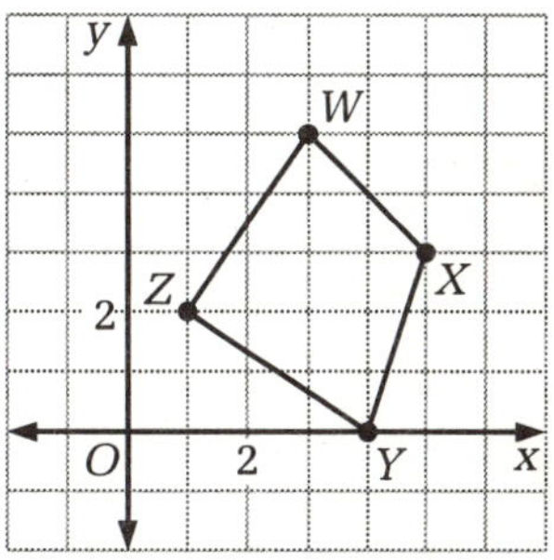

Sample Response

The orientation of quadrilateral $WXYZ$ is clockwise; the orientation of quadrilateral $W'X'Y'Z'$ is counterclockwise.

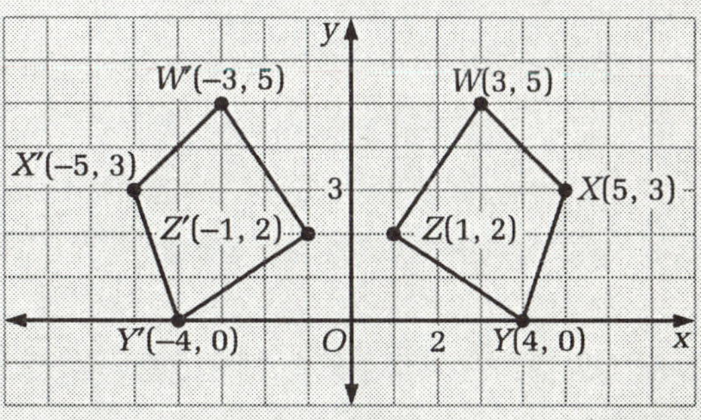

Copy each figure and reflect it over the indicated line. Describe the orientation of the original figure and its image.

5. x-axis

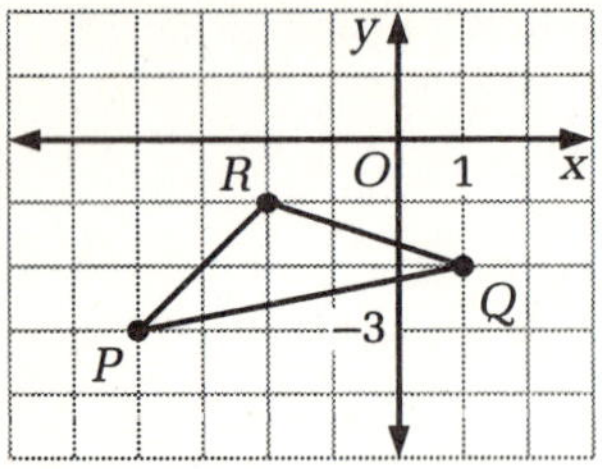

6. y-axis

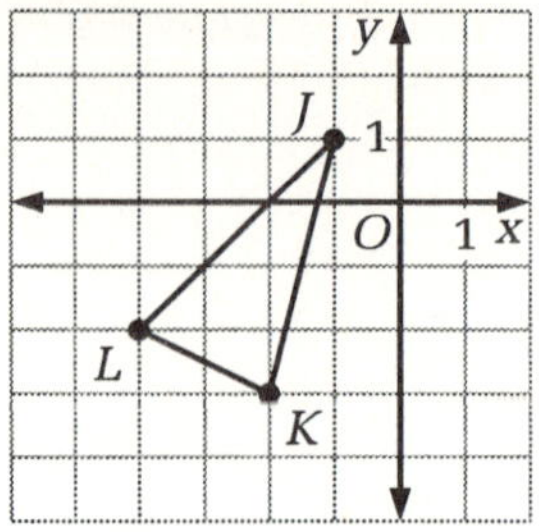

7. y-axis

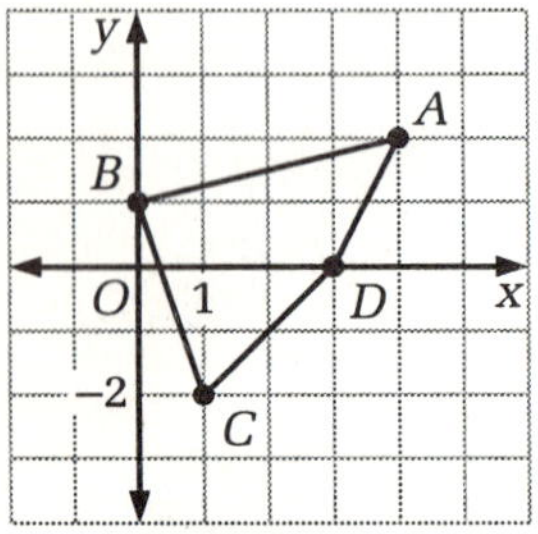

8. x-axis

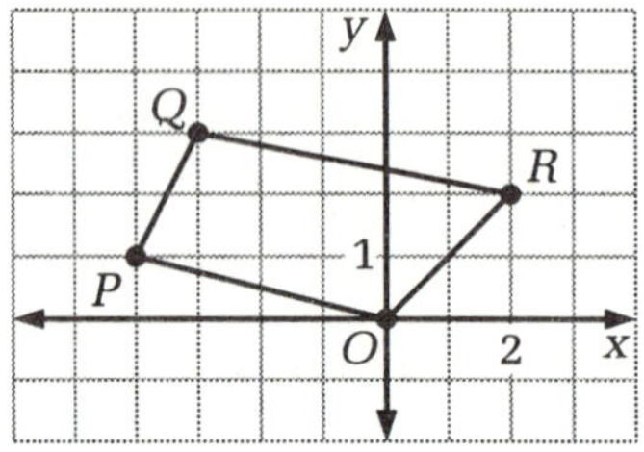

Review PREVIEW

9. Find the ratio of the volumes of a cone with radius 8 cm and a similar cone with radius 18 cm.

10. The ratio of the diameter of two similar cylinders is 5 : 2. The volume of the larger cylinder is 750π cm^3. What is the volume of the smaller cylinder?

Graph each equation.

11. $y = -2$ **12.** $x = 3$ **13.** $y = 1$ **14.** $x = -4$

Name ______________________________ Date ______________

Skills Inventory

Inventory I on Units 1–2

Answers

1. ____________
2. ____________
3. ____________
4. *See question.*
5. ____________
6. *See question.*
7. ____________

Directions: Write answers in the spaces provided.

1. A random sample of 120 students at Central High is polled and 45 are found to be in favor of extending the length of the school day by adding 15 minutes to the lunch break. How many of the entire student body of 1600 would be expected to favor this proposal?

2. It is estimated that 25% of the fish in a certain lake are trout. Which of the spinners below could be used to simulate the number of trout in a sample of 20 fish from the lake?

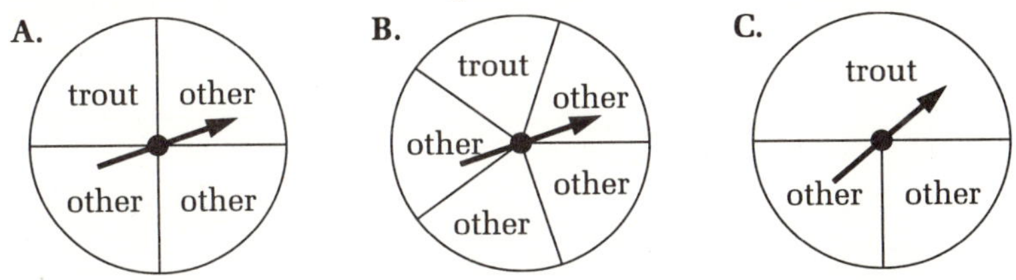

A group of workers for a survey company stands in the walkways of a mall and asks shoppers if they will participate in a survey. Those who respond "yes" are shown an advertisement for a new food product and asked a list of questions.

3. What type of sampling technique is this?

4. Describe one problem with this survey method.

Questions 5–7 refer to the pattern at the right.

$$1 + 3 = 4$$
$$1 + 3 + 5 = 9$$
$$1 + 3 + 5 + 7 = 16$$
$$1 + 3 + 5 + 7 + 9 = \_\_\_$$

5. Complete the last row of the pattern.

6. What conjecture can you make based on this series of equations?

7. Is this conjecture an example of inductive or deductive reasoning?

Name ______________________ Date ______________

Inventory I *(continued)*

8. Write the converse of the statement, "If a figure is a square then it has four right angles."

9. Is the converse you wrote for Exercise 9 *true* or *false*? If it is false, give a counterexample.

For Questions 10–13 refer to the graph at the right and describe the behavior of the function over each interval as increasing, decreasing, or constant and as linear or nonlinear.

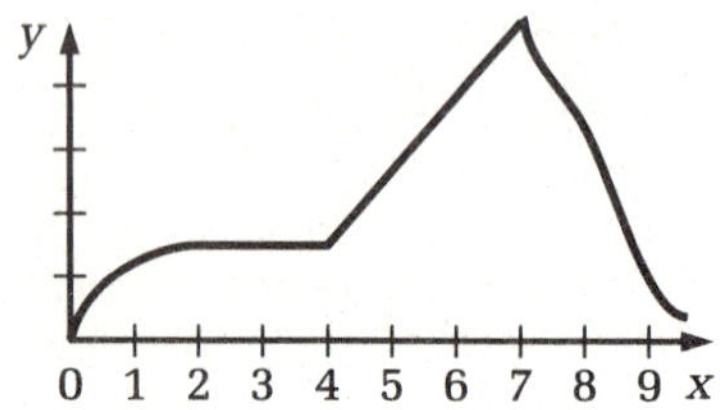

10. $0 < x < 2$ **11.** $2 < x < 4$ **12.** $4 < x < 7$ **13.** $x > 7$

14. Write an equation for the line passing through the points $(-4, 3)$ and $(-2, 1)$.

15. Is inverse variation a good model for this data? Why or why not?

x	1.000	1.333	1.500	1.750	1.800	1.950
y	2.000	1.500	1.333	1.143	1.111	1.026

16. The radius of a youth soccer ball is about 9.8 cm. Find its volume to the nearest cubic centimeter.

For Questions 17–19, find an equation that models each variation, using *k* as the variation constant.

17. The height, h, of a falling rock varies directly with the square of the time, t, since it was dropped.

18. The frequency of a wave, f, varies inversely with its wavelength, l.

19. The distance, D, traveled by a car varies directly with its average speed, r.

Simplify. Write answers with positive exponents.

20. $\frac{18x^{-2}}{2x^3}$ **21.** $\frac{(2 + x)^0}{x^{-2}}$ **22.** $49^{1/2}$ **23.** $(-27x)^{1/3}$

24. One isotope of radioactive sodium has a half-life of 1 minute. How much of a 30-g sample will remain after 10 min.?

Answers

8. *See question.*
9. *See question.*
10. ______
11. ______
12. ______
13. ______
14. ______
15. *See question.*
16. ______
17. ______
18. ______
19. ______
20. ______
21. ______
22. ______
23. ______
24. ______

Name ______________________________ Date ______________

Inventory 2 on Units 1–3

Answers

Directions: Write answers in the spaces provided.

1. A recent poll showed that 59 out of 100 voters were in favor of Proposition 9. How many of the city's 13,408 voters might be expected to vote for the proposition?

2. Each flower seed in a packet is said to have a $\frac{5}{6}$ chance of germinating when planted. Describe how you could use a die to simulate the results of planting 25 seeds.

An environmental group wishes to poll the population of a city about views on mosquito control. Classify each sampling method below as *random, convenience, stratified random, cluster*, or *systematic*. Explain your choice.

3. The group goes through the phone book and picks the first number on every tenth page.

4. The environmentalists set up a table next to a commuter train stop and ask each person who comes by.

5. A pair of researchers goes to each ward in the city and talks to the residents of one home on each block.

Lynn observed that $1^2 + 1 + 11 = 13$, which is prime; $2^2 + 2 + 11 = 17$, which is prime; and $3^2 + 3 + 11 = 23$, which is prime.

6. What conjecture might Lynn make based on this information?

7. Is this conjecture *true* or *false*? If false, find a counterexample.

8. Write the converse of the statement, "If a number is divisible by 6, then is it divisible by 2."

9. Write an equation for the line through (–4, 4) and (2, 0).

10. Jorge's kitten has two toy balls. One has a circumference of 9 cm and the other has a circumference of 6 cm. If the larger ball has a surface area of about 25.8 cm^2, what is the surface area of the smaller one?

1. ______________
2. *See question.*
3. ______________
4. ______________
5. ______________
6. *See question.*
7. *See question.*
8. *See question.*
9. ______________
10. ______________

Name ______________________ Date ____________

Inventory 2 *(continued)*

For questions 11–13, write an equation that models the variation with variation constant k.

11. The volume of a sphere, V, is directly proportional to the cube of its radius, r.

12. Among rectangles having a fixed area, the length, l, varies inversely with its width, w.

13. The strength of a bone, s, varies directly with the square of its cross-sectional radius, r.

Simplify. Write answers with positive exponents.

14. $\dfrac{(-5x^0)^3}{10x^2}$ 15. $64^{1/3}$ 16. $\dfrac{3ab}{b^{-2}}$ 17. $\sqrt{4r^2t^4}$

18. Identify the system of inequalities that is graphed at the right.

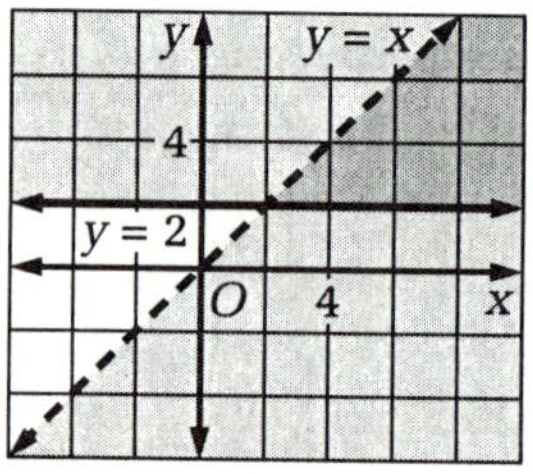

Find the solution of each system of equations.

19. $y = 4x$
$y = 4x - 5$

20. $y = x - 7$
$y = 3x + 5$

21. $2x + 3y = 3$
$2x - 5y = 11$

Perform each indicated matrix operation.

22. $\begin{bmatrix} 2 & 3 & 1 \\ 4 & 1 & -2 \end{bmatrix} - 2\begin{bmatrix} 1 & -2 & 4 \\ 0 & 5 & 0 \end{bmatrix}$

23. $\begin{bmatrix} 2 & 1 \\ 0 & 2 \\ 3 & -3 \end{bmatrix}\begin{bmatrix} -1 & 4 \\ 0 & -2 \end{bmatrix}$

Describe the transformation represented by each matrix operation.

24. $0.5\begin{bmatrix} x \\ y \end{bmatrix}$

25. $\begin{bmatrix} x \\ y \end{bmatrix} + \begin{bmatrix} -2 \\ 1 \end{bmatrix}$

Answers

11. ____________
12. ____________
13. ____________
14. ____________
15. ____________
16. ____________
17. ____________
18. ____________
19. ____________
20. ____________
21. ____________
22. *See question.*
23. *See question.*
24. *See question.*
25. *See question.*

Name ______________________ Date ____________

Inventory 3 on Units 1–4

Answers

1. ____________
2. *See question.*
3. ____________
4. *See question.*
5. ____________
6. *See question.*
7. ____________

Directions: Write answers in the spaces provided.

1. A random sample of 25 pages of a novel was carefully examined, and 3 misspellings were found. About how many misspellings might be expected in the entire 327-page book?

2. It is estimated that $\frac{1}{3}$ of the people who call for a doctor's appointment because of illness have some form of the common cold. Mark the spinner at the right so that it could be used to simulate a sample from the population of calls to the doctor.

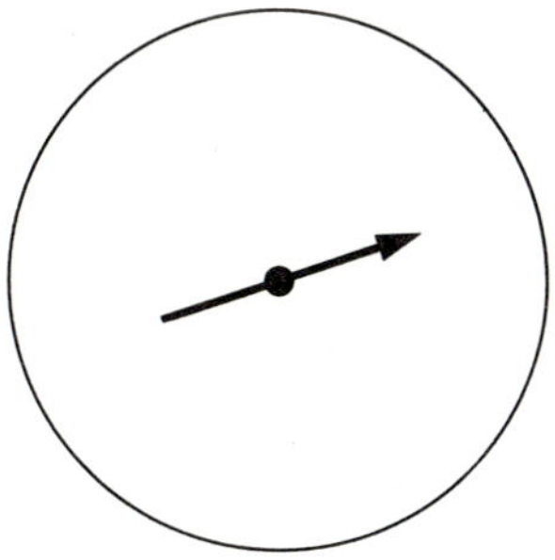

A radio station sets up a call-in line for people to register their opinion about whether the verdict in a recent well-publicized trial was fair. The call-in line is in operation from 5 P.M. to midnight for 3 days. The results of the survey are reported on the air.

3. What type of sampling method is this: *random, convenience, stratified random, cluster* or *systematic*?

4. Describe one problem with this sampling method.

Observe the following equations: $1 + 2 = 3 = \frac{2 \cdot 3}{2}$, $1 + 2 + 3 = 6 = \frac{3 \cdot 4}{2}$, and $1 + 2 + 3 + 4 = 10 = \frac{4 \cdot 5}{2}$.

5. Predict what $1 + 2 + 3 + \ldots + 8$ will equal.

6. What conjecture can you make based on the pattern shown by these equations?

7. Is this conjecture an example of inductive or deductive reasoning?

Name ______________________ Date __________

Inventory 3 *(continued)*

Consider the following pair of implications.
If you live in Los Angeles, then you live in California.
If you live in California, then you own a car.

8. Which of the Venn diagrams below correctly represents these two statements?

A.

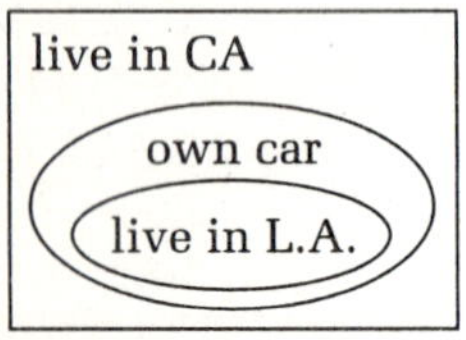

B.

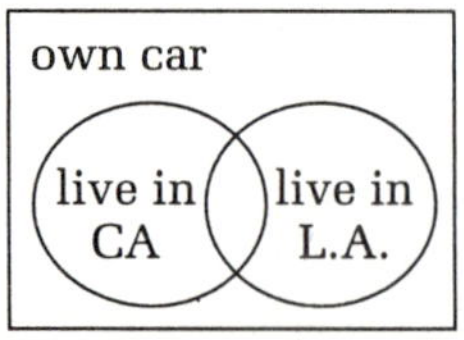

C.

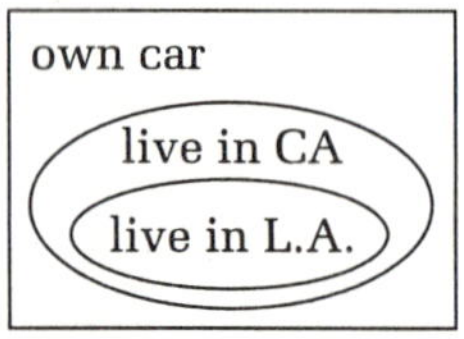

9. What is the converse of the second implication?

This graph shows the amount of sales tax on an item costing up to $1.00.

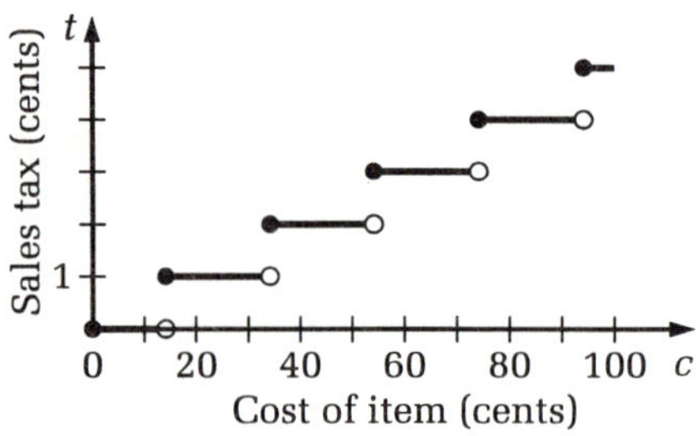

10. What is the domain of this function?

11. What is the range of this function?

12. Find the slope of the line connecting (0, 5) and (–3, 6).

13. Which of the data pairs below satisfy the relation y varies inversely with x with variation constant 5?

A. (1, 5) **B.** (5, 5) **C.** (–1, –5) **D.** $\left(10, \frac{1}{2}\right)$ **E.** $\left(\frac{5}{2}, 2\right)$

14. A spherical balloon with a diameter of 10 cm has a volume of about 524 cm^3. If more air is added to the balloon so that its diameter expands to 20 cm, what is its new volume?

The distance a rock falls, *d*, varies directly with the square of the time, *t*, after its release. If *d* is measured in meters and *t* in seconds, the constant of variation is 4.9.

15. Write an equation that shows the relationship between d and t.

16. How far does the rock fall in the first 3 seconds after its release?

Answers

8. ______________
9. *See question.*
10. *See question.*
11. *See question.*
12. ______________
13. ______________
14. ______________
15. ______________
16. ______________

Name ______________________ Date ______________

Inventory 3 *(continued)*

Simplify. Write answers with positive exponents.

17. $\frac{7x^0}{(14x)^2}$ **18.** $\frac{6a^2b^{-3}}{4a^{-1}b}$ **19.** $-125^{1/3}$ **20.** $(49x)^{1/2}\sqrt{xy}$

21. A colony of bacteria is growing in a laboratory. At 10:00 A.M. there were 20 bacteria in the colony and at 11:00 A.M. there were 40 bacteria. If this rate of growth is sustained, how many bacteria will there be at 5:00 P.M.?

22. What is the solution of the system of equations graphed at the right?

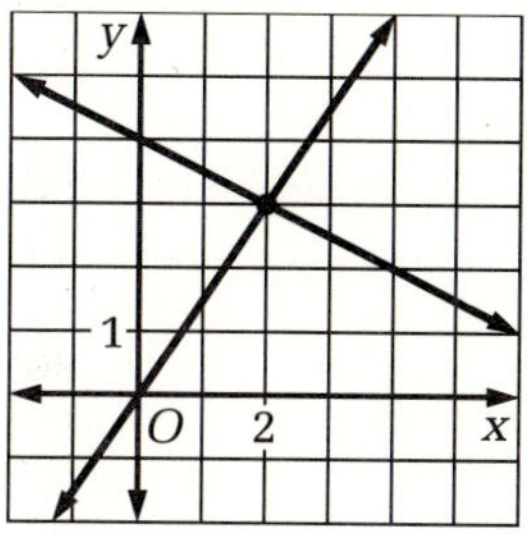

Find the solution of each system of equations.

23. $y = -x$
$y = 2x - 21$

24. $2x + y = 5$
$2x + y = -3$

25. $3x + 2y = 3$
$2x - 3y = -11$

26. What is the slope of a line perpendicular to $5x - 2y = 4$?

For exercises 27 and 28 let $A = \begin{bmatrix} 4 & 2 \\ -3 & 0 \end{bmatrix}$ and $B = \begin{bmatrix} -7 & 1 \\ 2 & -2 \end{bmatrix}$.

Perform each matrix operation.

27. $A + B$ **28.** $2A - 3B$

Exercises 29 and 30 refer to $\triangle ABC$, with $A(2, -2)$, $B(4, 4)$ and $C(3, -6)$.

29. Write a matrix expression that represents dilation of this triangle with scale factor $\frac{3}{2}$ centered at the origin.

30. Write a matrix representing the results of this dilation.

31. Describe the transformation represented by the matrix expression $\begin{bmatrix} x \\ y \end{bmatrix} + \begin{bmatrix} 4 \\ -5 \end{bmatrix}$.

Answers

17. ______________

18. ______________

19. ______________

20. ______________

21. ______________

22. ______________

23. ______________

24. ______________

25. ______________

26. ______________

27. *See question.*

28. *See question.*

29. *See question.*

30. *See question.*

31. *See question.*

Name ______________________ Date ______________

Inventory 3 *(continued)*

Find each product.

32. $\begin{bmatrix} 2 & 3 \\ 0 & -4 \end{bmatrix}\begin{bmatrix} 1 & -2 \\ -3 & -1 \end{bmatrix}$

33. $\begin{bmatrix} 3 & 0 & 5 \end{bmatrix}\begin{bmatrix} 2 & 1 & 6 \\ 1 & -4 & -3 \\ 4 & 0 & -3 \end{bmatrix}$

34. Write a matrix equation equivalent to this system of equations.
$x - 3y = 12$
$4x + y = 9$

35. Find the vertex of the parabola whose equation is $y = -3x^2 + 12x - 1$.

For Exercises 36–40,match the letter of the graph to its equation.

36. $y = x^2 - 4$

37. $y = (x - 2)^2$

38. $y = -x^2 + 4$

39. $y = -(x + 1)^2$

40. $y = 2x^2 - 4$

A.
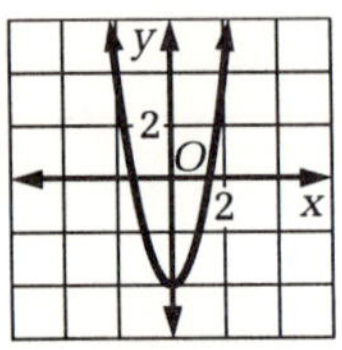

B.
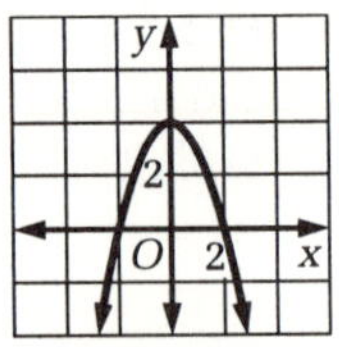

C.
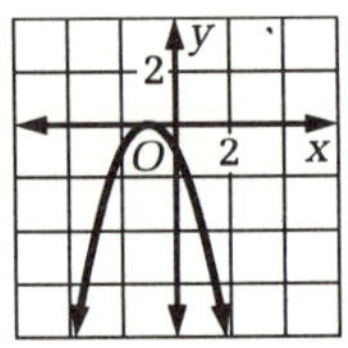

D.
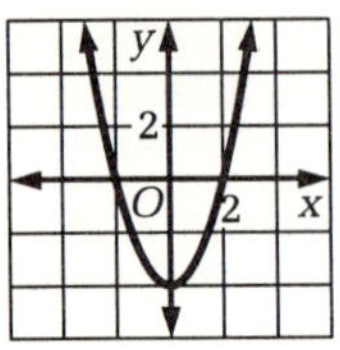

E.
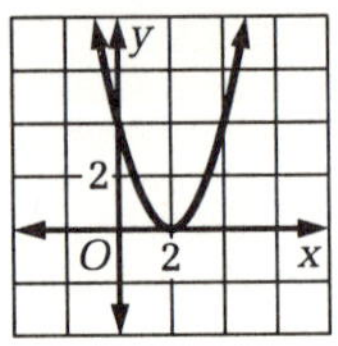

Solve each equation.

41. $2x^2 - 6 = 12$ **42.** $x^2 - 4x = 5$ **43.** $4(x - 1)^2 - 9 = 0$

44. Factor completely: $x^4 - 81$.

45. Solve $2x^2 + 5x - 1 = 0$, giving solutions to the nearest hundredth.

For exercises 46–48, use the discriminant to classify the roots of each equation.

46. $2x^2 + 3x - 1 = 0$ **47.** $2x^2 + 8x + 8 = 0$ **48.** $2x^2 + 8x + 9 = 0$

Simplify.

49. $\sqrt{-49}$ **50.** $(-2 - 3i)(1 + i)$

Answers

32. *See question.*
33. *See question.*
34. *See question.*
35. ______
36. ______
37. ______
38. ______
39. ______
40. ______
41. ______
42. ______
43. ______
44. *See question.*
45. ______
46. ______
47. ______
48. ______
49. ______
50. ______

Name ______________________ Date ____________

Inventory 4 on Units 1–5

Directions: Write answers in the spaces provided.

1. A quality control department tested a random sample of 100 toasters as they came off the assembly line and found 2 to be defective. About how many of the day's output of 5,080 toasters might be expected to be defective?

2. It is estimated that one out of 8 pairs of nesting herons successfully raises two chicks in any particular year. The other nesting pairs raise fewer than 2 chicks each. Describe how the spinner at the right could be used to simulate a random sample of 20 such nesting pairs.

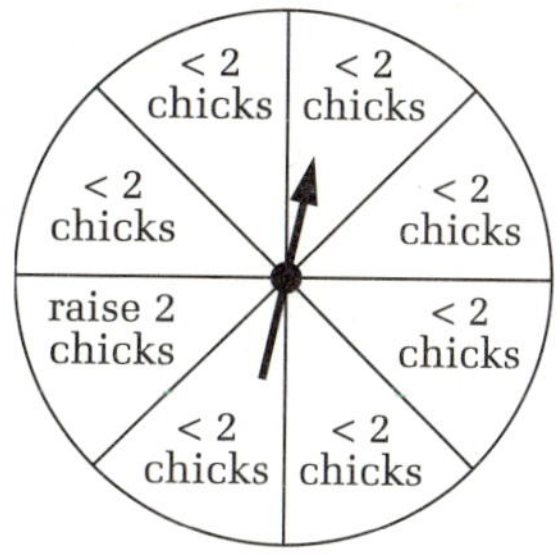

A grocery store owner wished to assess how many of her customers would come in to shop if her store was open from noon to 6 P.M. on an upcoming holiday. Classify each sampling method described below as *random, convenience, stratified random, cluster*, or *systematic*.

3. She stands in a randomly selected aisle and asks each shopper who comes down that aisle.

4. She gives a stack of ballots to each cashier to be handed out to each customer who buys groceries.

5. Describe one problem with the sampling method given in Exercise 4.

Chen drew 10 triangles and carefully measured their angles. He noted that the sum of the angle measures for each triangle was about 180°.

6. What conjecture might Chen make about the angles of a triangle?

7. Is this conjecture an example of inductive or deductive reasoning?

8. Write the converse of the statement, "If a quadrilateral has four right angles, then it is a rectangle."

9. Is the converse you wrote for Exercise 8 *true* or *false*? If false, give a counterexample.

Answers

1. ____________
2. *See question.*
3. ____________
4. ____________
5. *See question.*
6. *See question.*
7. ____________
8. *See question.*
9. *See question.*

Name ____________________ Date ____________

Inventory 4 *(continued)*

10. At a recent board meeting the circulation manager of the local library gave this report: "Circulation decreased gradually over the first 5 months of the year, stayed fairly constant over the summer months, and then increased rapidly from September to the start of the holiday season, when it again declined." Which graph below best represents circulation over this one-year period?

A.

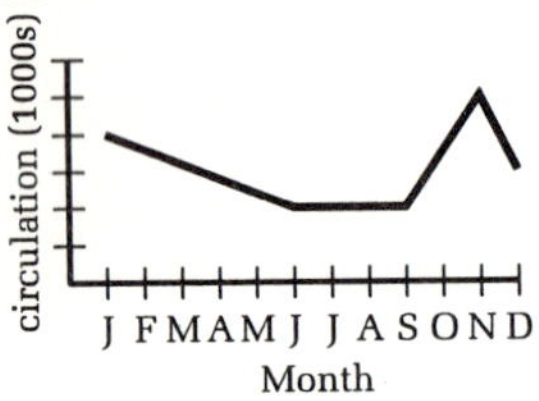

B.

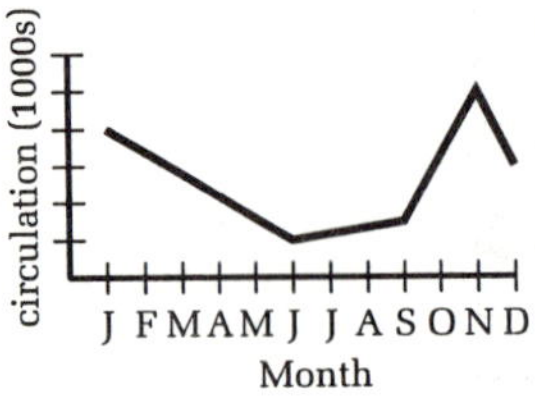

C.

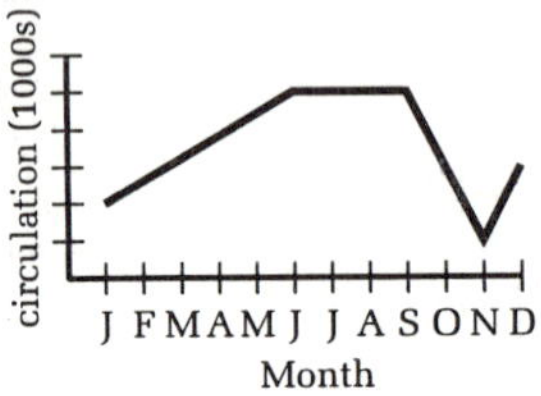

11. Find the slope of the line graphed at the right.

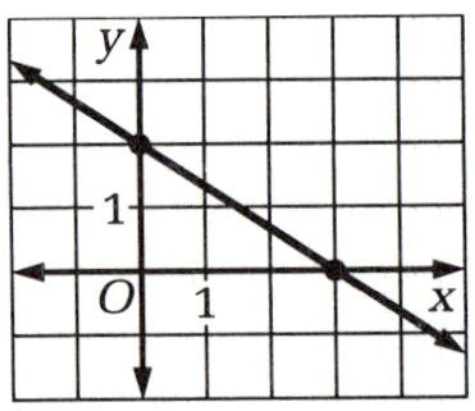

12. Complete the table of x and y values below given that y varies inversely with x with variation constant 3.

x	2.0	1.5		0.75
y			3	

13. The diameter of a golf ball is about 4.3 cm. Find its volume to the nearest tenth of a cubic centimeter.

14. Write an equation for y as a function of x if y varies directly with the square of x, and $y = 2$ when $x = 4$.

Simplify. Write answers with positive exponents.

15. $6a^0b^{-3}$ **16.** $\frac{10x^4}{25y^{-2}}$ **17.** $81^{1/2}$ **18.** $(-8xy^3)^{1/3}$

19. At an interest rate of 4.5% an investment of \$100 doubles to \$200 in about 15.75 years. At this rate how long will it take for the investment to reach \$1600?

20. What system of inequalities is at the right?

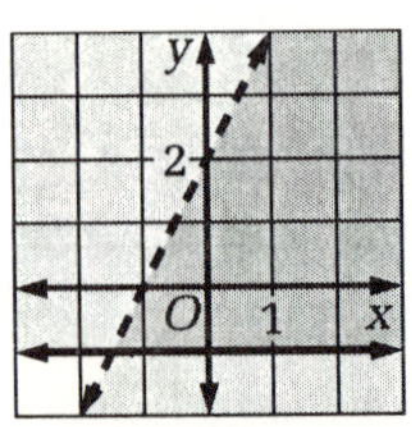

Answers

10. ____________

11. ____________

12. *See question.*

13. ____________

14. ____________

15. ____________

16. ____________

17. ____________

18. ____________

19. ____________

20. *See question.*

Name ______________________ Date __________

Inventory 4 *(continued)*

Solve each system of equations.

21. $y = 4x - 5$
$y = 2x + 1$

22. $2x + 3y = 5$
$3x - 2y = -12$

23. Write an equation for the line through the point (2, 3) perpendicular to the line $2x + y = 4$.

Perform each matrix operation.

24. $\begin{bmatrix} 3 & -2 & 1 \\ 1 & 0 & 4 \end{bmatrix} + \begin{bmatrix} -1 & 1 & 5 \\ -7 & -2 & 3 \end{bmatrix}$

25. $\frac{1}{2}\begin{bmatrix} 3 & -5 \\ 2 & 3 \end{bmatrix}$

Describe the transformation indicated by each matrix operation.

26. $\begin{bmatrix} x \\ y \end{bmatrix} + \begin{bmatrix} -2 \\ -3 \end{bmatrix}$

27. $\frac{2}{3}\begin{bmatrix} x \\ y \end{bmatrix}$

28. Find the product $\begin{bmatrix} 4 & -2 & -3 \\ 0 & 1 & 7 \end{bmatrix}\begin{bmatrix} 10 & 2 \\ 2 & -9 \\ 0 & 3 \end{bmatrix}$.

29. Given that the inverse of the matrix $\begin{bmatrix} 2 & 1 \\ 5 & 3 \end{bmatrix}$ is $\begin{bmatrix} 3 & -1 \\ -5 & 2 \end{bmatrix}$, write the matrix equation that represents the solution of $\begin{bmatrix} 2 & 1 \\ 5 & 3 \end{bmatrix}\begin{bmatrix} x \\ y \end{bmatrix} = \begin{bmatrix} 1 \\ 3 \end{bmatrix}$.

30. Write the solution to the matrix equation in Exercise 29.

31. Find the vertex of the parabola $y = x^2 - 3x - 1$.

Each of the functions graphed below is a transformation of the parabola $y = x^2$. Write an equation for each function.

32.

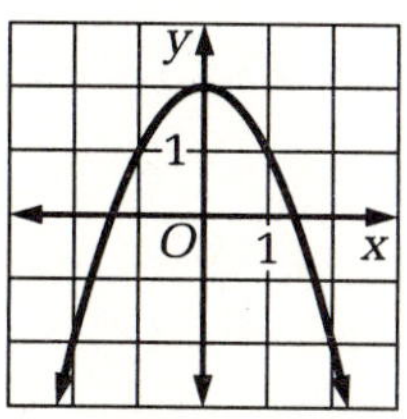

33.

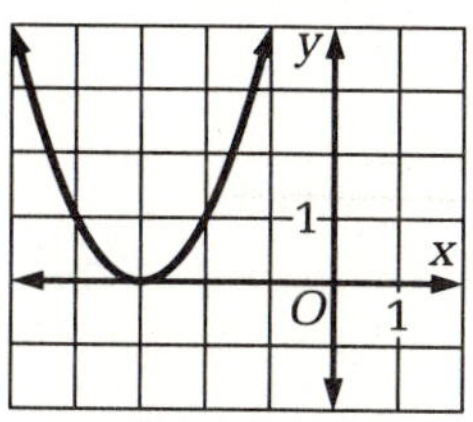

34.

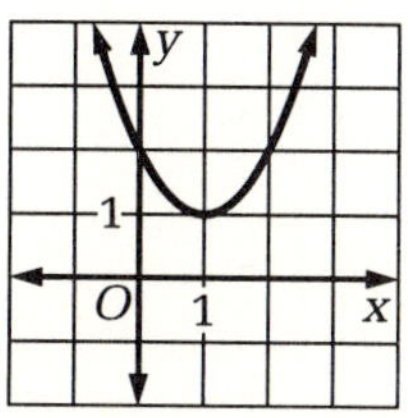

35.

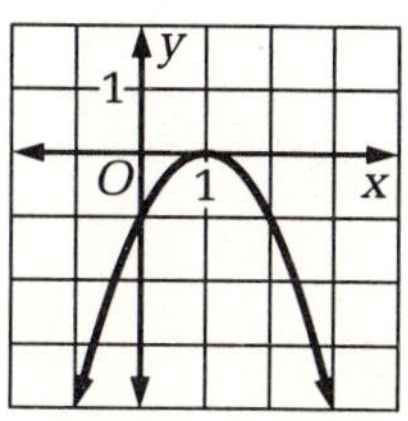

Answers

21. ______________
22. ______________
23. ______________
24. *See question.*
25. *See question.*
26. *See question.*
27. *See question.*
28. *See question.*
29. *See question.*
30. *See question.*
31. ______________
32. ______________
33. ______________
34. ______________
35. ______________

Name ______________________ Date ____________

Inventory 4 *(continued)*

Answers

36. Factor completely: $3x^3 - 6x^2 - 24x$.

Solve each quadratic equation.

37. $(x + 1)^2 - 3 = 6$

38. $9x^2 + 6x + 1 = 0$

39. Solve $3x^2 - 4x - 2 = 0$, giving solutions to the nearest hundredth.

For Exercises 40–42, use the discriminant to classify the roots of each equation.

40. $3x^2 - 4x + 1 = 0$ **41.** $x^2 - 4x + 7 = 0$ **42.** $x^2 - 4x + 4 = 0$

43. Find all solutions of the quadratic system: $y = -x^2 + 2x + 3$
$y = x^2 - 2x - 3$

44. Draw a Venn diagram that shows the relationships among the sets of parallelograms, rhombuses and squares.

Exercises 45–48 refer to quadrilateral *ABCD* with vertices *A*(0, 0), *B*(1, 2), *C*(3.5, 1) and *D*(2.5, –1).

45. Use the slopes of lines *AB, BC, CD,* and *DA* to show that *ABCD* is a parallelogram.

46. Is quadrilateral *ABCD* a rhombus? How do you know?

47. Find the coordinates of the midpoint of diagonal *AC*.

48. Find the coordinates of the vertices of the image obtained by reflecting *ABCD* over the *x*-axis.

49. Give the coordinates of each vertex of the kite drawn at the right.

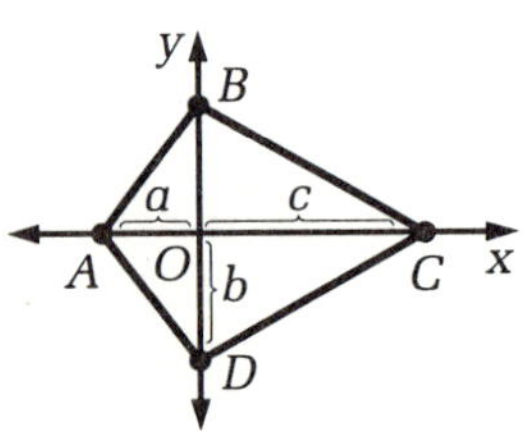

50. Name one property of the diagonals of a kite that is evident from the graph.

36. ____________
37. ____________
38. ____________
39. ____________
40. ____________
41. ____________
42. ____________
43. ____________
44. *See question.*
45. *See question.*
46. *See question.*
47. ____________
48. *See question.*
49. *See question.*
50. *See question.*

Name ______________________________ Date ______________

Inventory 5 on Units 6–7

Answers

1. *See question.*
2. ______
3. ______
4. ______
5. *See question.*
6. ______
7. ______
8. ______
9. ______
10. *See question.*
11. *See question.*

Directions: Write answers in the spaces provided.

1. Make a tree diagram showing the possible combinations of heads and tails in three flips of a coin.

2. Jeanette is packing for a business trip. She packs 2 blazers, 3 blouses and 3 skirts that all coordinate. How many different outfits of blazer, blouse, and skirt can she wear?

3. 34% of the registered voters in Smallville are registered as Republicans, and 29% are registered as independents. What percentage are registered as Democrats?

4. If the odds against winning a door prize are 5 : 1, what is the probability of winning one?

5. Suppose A and B are events such that $P(A) = \frac{1}{2}$, $P(B) = \frac{1}{3}$, and $P(A \text{ and } B) = \frac{1}{5}$. Are A and B independent? How do you know?

6. A box contains 20 tokens, 10 blue and 10 red. Two are drawn at random. What is the probability both are blue?

7. The summer reading list includes 10 books, of which students must read at least 3. How many different sets of 3 books can be selected?

8. What is the sum of the elements of row 8 of Pascal's triangle?

9. Kendra has been hitting 70% of her free throws in the last four basketball games. What is the probability she will make 4 out of her next 5 free throws?

10. Write the expanded form of $(x + y)^4$.

11. What conclusion can you reach from the given conditions: Either Jim or Tomas called while I was out. Jim is on a camping trip.

Name ______________________ Date __________

Inventory 5 *(continued)*

Exercises 12 and 13 refer to the implication "$x^2 = 9$ only if $x = 3$."

12. Rewrite the implication in if-then form.

13. Is the implication true or false? If it is false, give a counterexample.

Identify the rule of logic used in each valid argument.

14. Tsuruko is in the computer room or she is in the chemistry lab. Tsuruko is not in the chemistry lab. Tsuruko must be in the computer room.

15. If today is Wednesday, then we are in Belgium. We are not in Belgium. So, today is not Wednesday.

Tell whether each biconditional is true or false. Explain your reasoning.

16. Tomorrow is Friday if and only if today is Thursday.

17. A triangle is acute if and only if it has an acute angle.

18. Write a flow proof of "If $3x - 5 = 1$, then $x = 2$."

19. Write a two-column proof of: If $\angle 1$ and $\angle 2$ are supplementary angles and $\angle 1$ is a right angle, then $\angle 2$ is a right angle.

Exercises 20 and 21 use the figure at the right, where lines *l* and *m* are parallel and are intersected by transversal *t*.

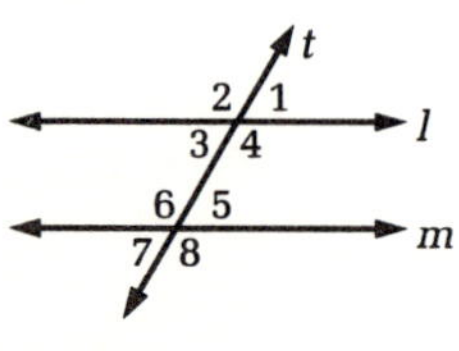

20. List the pairs of corresponding angles.

21. If $m\angle 1 = 42°$, what is $m\angle 2$? $m\angle 4$? $m\angle 5$? $m\angle 8$?

Answers

12. *See question.*

13. *See question.*

14. ______

15. ______

16. *See question.*

17. *See question.*

18. *See question.*

19. *See question.*

20. *See question.*

21. ______

Name ______________________________ Date ______________

Inventory 6 on Units 6–8

Answers

1. *See question.*
2. ____________
3. ____________
4. ____________
5. ____________
6. ____________
7. ____________
8. ____________
9. *See question.*
10. *See question.*

Directions: Write answers in the spaces provided.

1. A string quartet will play works by Mozart, Brahms and Hindemith at a concert. Draw a tree diagram showing how many different ways the three concert pieces can be arranged in the program.

2. How many different 4-letter arrangements can be made from the letters in the word SPRING? Assume letters can be used only once.

3. When Jasmine comes up to bat against the Hornets, there is a 20% chance that she will strike out and a 35% chance she will make some other kind of out. What is the probability she will get on base?

A regular deck of 52 playing cards contains 26 black and 26 red cards. Two are drawn at random. What is the probability that both are red if:

4. They are drawn together without replacement?

5. The first is returned to the deck, and the deck is reshuffled before the second is drawn?

6. In the game of cribbage, each player is dealt 6 cards from a standard 52 card deck. How many different cribbage hands are possible?

7. A packet of zucchini squash seeds claims a germination rate of 90%. If 4 seeds are planted, what is the probability that at least 3 will germinate?

8. Write out the fourth row of Pascal's triangle.

9. Find the expanded form of $(a - 2)^4$.

10. Graph the disjunction $x < 1$ or $x > 5$ on a number line.

Name ______________________ Date ____________

Inventory 6 *(continued)*

Answers

11. Write the implication, "Raul goes sailing whenever he gets a day off work." in if-then form.

12. What conclusion can be reached from the following premises:
If I want to buy a new swimsuit, I must go to the mall.
To go to the mall, I must take the bus.
I really want to get a new swimsuit.

13. Rewrite the definition below as a biconditional.
A right triangle has one right angle.

14. Write a flow proof of "If $7x + 3 = -4$, then $x = -1$."

15. Given that lines l and m are intersected by line t and $m\angle 1 = m\angle 3$, write a two-column proof to show that $m\angle 2 = m\angle 3$.

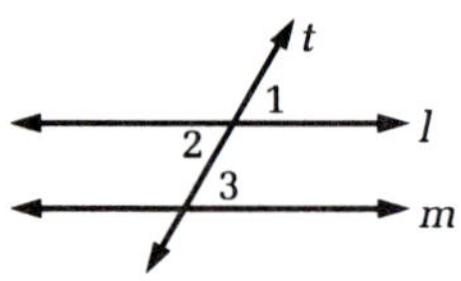

16. Write a paragraph proof to show that in parallelogram $ABCD$, $\angle A$ and $\angle B$ are supplementary.

11. *See question.*
12. *See question.*
13. *See question.*
14. *See question.*
15. *See question.*
16. *See question.*

Name ______________________________ Date ____________

Inventory 6 *(continued)*

Tell whether lines l_1 and l_2 are parallel in each case, given the measures of $\angle 1$ and $\angle 2$.

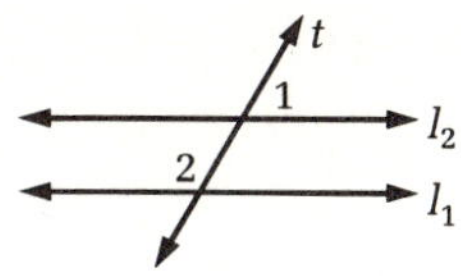

17. $m\angle 1 = 80°$, $m\angle 2 = 80°$

18. $m\angle 1 = 49°$, $m\angle 2 = 131°$

19. Find the measure of each angle of $\triangle ABC$.

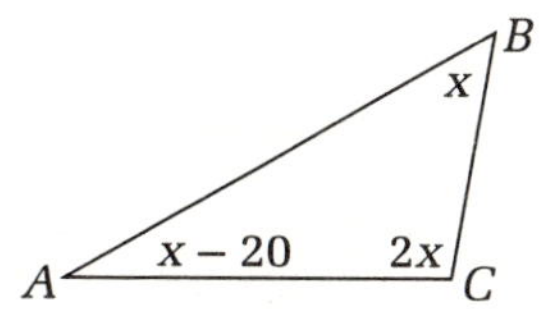

Use the diagram at the right for Exercises 20 and 21.

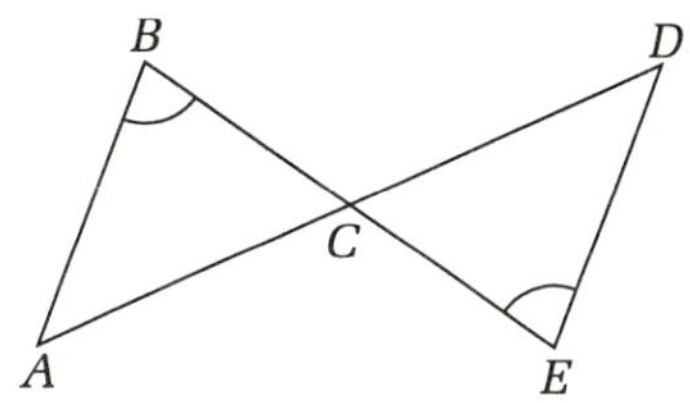

20. In the diagram, how do you know that $\triangle ABC \sim \triangle DEC$?

21. List two possible sets of conditions that could be used to prove $\triangle ABC \cong \triangle DEC$, along with the corresponding theorem or postulate.

22. To draw the bisector of $\angle A$, Kim marked points B and C, equidistant from A, and then found point D, in the interior of $\angle A$, equidistant from B and C. Prove that $\overrightarrow{AD}$ is indeed the bisector of $\angle A$.

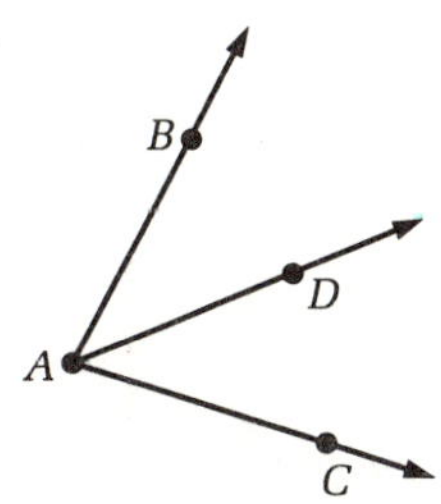

23. Identify the similar triangles in the figure at the right.

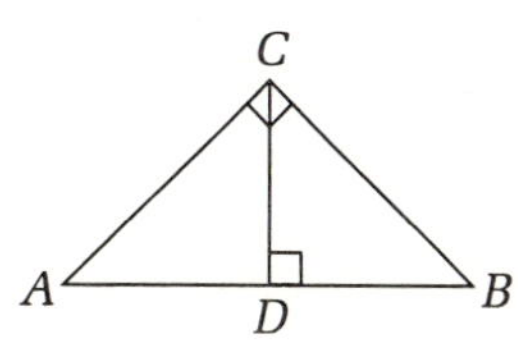

Answers

17. ____________

18. ____________

19. *See question.*

20. *See question.*

21. *See question.*

22. *See question.*

23. *See question.*

Name ______________________ Date ____________

Inventory 6 *(continued)*

Answers

Find the value of the variable(s) in each triangle.

24.

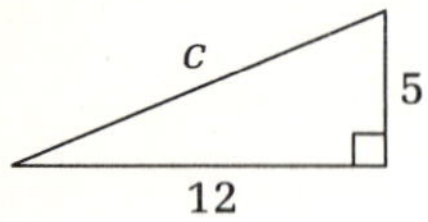

25.

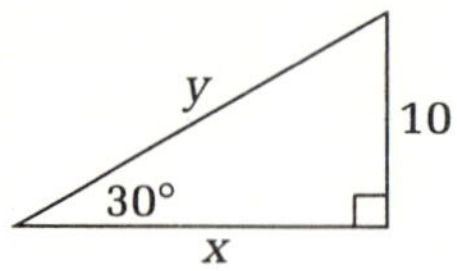

26.

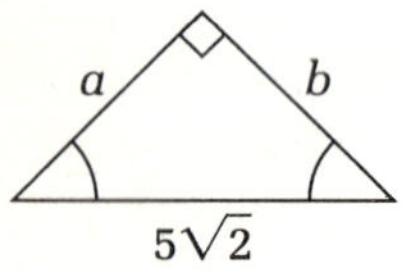

24. ____________

25. ____________

26. ____________

Name ______________________________ Date ______________

Inventory 7 on Units 6–9

Directions: Write answers in the spaces provided.

1. Melanie's three favorite flavors of ice cream are mint chocolate chip, heavenly hash, and French vanilla. Draw a tree diagram that shows the different orders in which these three flavors can be stacked to make a triple cone.

2. How many different 5-letter arrangements can be made from the letters in the word COUNT? Assume letters can be used only once.

Five red, four blue, three green and two yellow cards are mixed up together in a hat. One card is drawn at random.

3. What is the probability that it is not red?

4. What are the odds in favor of drawing a green card?

5. Suppose A and B are independent events with $P(A) = \frac{1}{4}$ and $P(B) = \frac{1}{5}$. What is $P(A \text{ and } B)$?

6. 48% of the voters in a certain district are Democrats, and 60% of the Democrats in the city are in favor of Proposition B. What is the probability that a randomly selected voter from this district is a Democrat in favor of Proposition B?

7. A team of 17 soccer players is forming a committee to plan an end of the season party. How many different 5-member committees are possible?

8. Row 5 of Pascal's triangle is 1 5 10 10 5 1. List the elements of row 6.

9. About 15% of Americans have Rh^- blood. What is the probability that exactly 2 of a random sample of 8 Americans are Rh^-?

10. Expand $(x + 1)^6$.

11. Every time Rosa goes to the store she buys potato chips or pretzels. Today Rosa went to the store and didn't buy any pretzels. What conclusion can be drawn?

12. Write an if-then statement corresponding to the Venn diagram at the right.

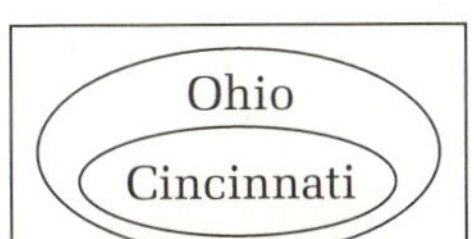

Answers

1. *See question.*
2. ______
3. ______
4. ______
5. ______
6. ______
7. ______
8. ______
9. ______
10. *See question.*
11. *See question.*
12. *See question.*

Name ______________________ Date ______________

Inventory 7 *(continued)*

In Exercises 13–15, what conclusion can be reached if both premises given are true?

13. A spider is an arachnid or an insect.
A spider is not an insect.

14. If $2x + 4 = 8$, then $2x = 4$.
If $2x = 4$, then $x = 2$.

15. If it is Consuela's birthday, then she will have a party.
Consuela is not having a party.

16. Identify the two conditionals in the biconditional, "A quadrilateral is a rectangle if and only if it has four right angles."

17. Write a flow proof of: If $4x^2 = 36$ and $x < 0$, then $x = -3$.

18. Given lines l and m intersected by line t, as shown in the diagram, with $m\angle 1 = m\angle 2$. Write a two-column proof to show that $\angle 3$ and $\angle 2$ are supplementary.

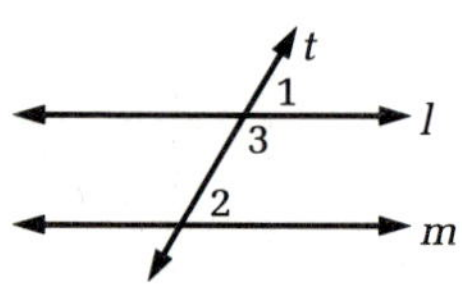

Quadrilateral *ABCD* is a parallelogram.

19. Name two pairs of parallel lines.

20. If $m\angle A = x°$, what is $m\angle B$? $m\angle C$?

21. What is $m\angle A + m\angle B + m\angle C + m\angle D$?

Name the theorem or postulate that tells you lines *l* and *m* are parallel in each case below.

22.

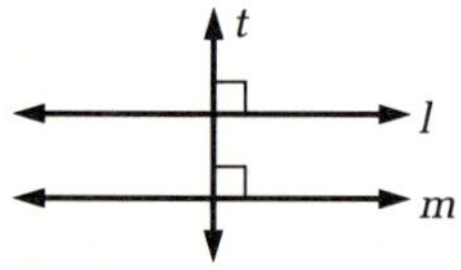

23.

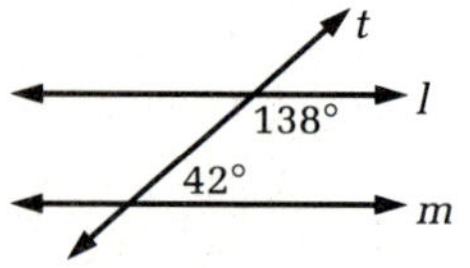

Answers

13. *See question.*
14. *See question.*
15. *See question.*
16. *See question.*
17. *See question.*
18. *See question.*
19. ______
20. ______
21. ______
22. *See question.*
23. *See question.*

Name ______________________________ Date ____________

Inventory 7 *(continued)*

In Exercises 24–26, find the unknown angle measures.

24.

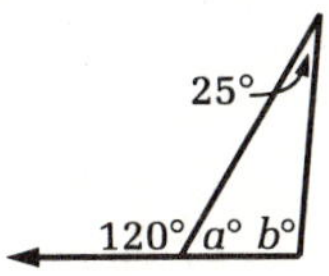

25.

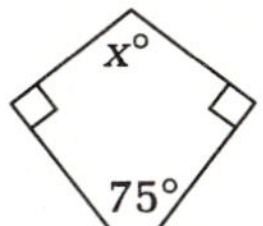

26.

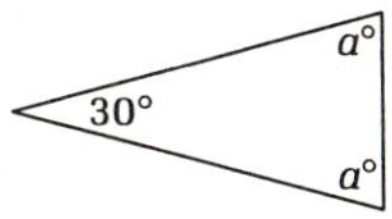

27. Find the lengths of $\overline{AC}$ and $\overline{BD}$.

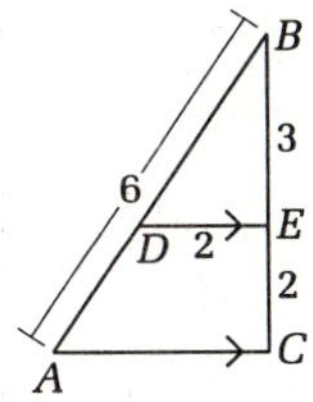

In each figure, identify a pair of congruent triangles, and tell how you know they are congruent.

28.

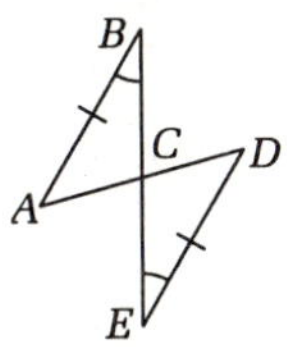

29.

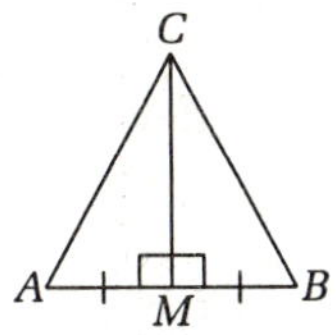

30. $\triangle ABC$ is isosceles, and $\overline{BD}$ bisects $\angle B$. How are $\overline{BD}$ and $\overline{AC}$ related?

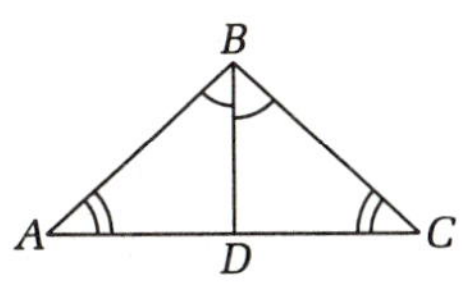

Find the exact value of each variable.

31.

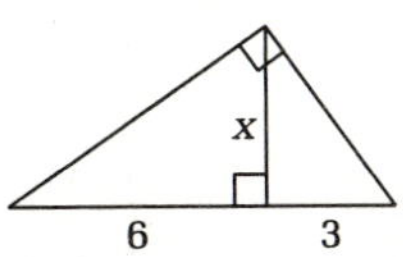

32.

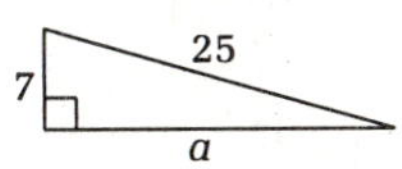

33.

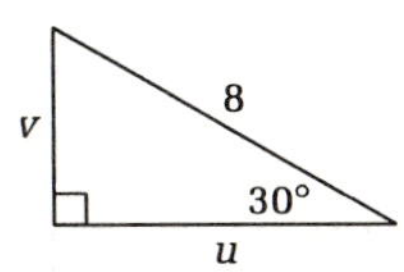

34. Give the values of sine, cosine, and tangent for $\angle ABC$.

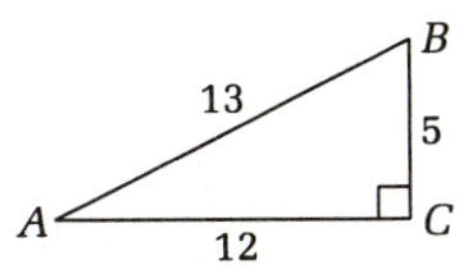

35. Write an equation to model the relationship between the surface area and dimensions of this cylinder.

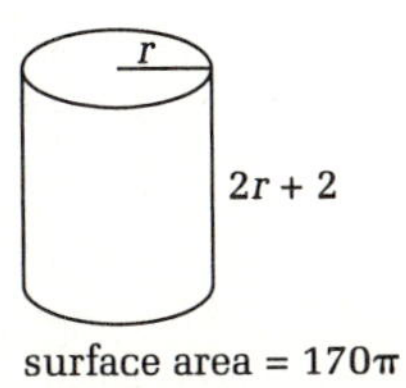

Answers

24. ____________

25. ____________

26. ____________

27. ____________

28. *See question.*

29. *See question.*

30. *See question.*

31. ____________

32. ____________

33. ____________

34. *See question.*

35. ____________

Name ________________________ Date ____________

Inventory 7 *(continued)*

Exercises 36–38 refer to the polynomial $4x^2 - 5x^3 + 12 + x^4$.

36. What is the degree of this polynomial?

37. Rewrite the polynomial in standard form.

38. What is the degree of the fourth term?

Simplify. Write without negative exponents.

39. $(-4x^2y)^3$

40. $\left(\dfrac{6a^3b}{4a^2b^2}\right)^{-1}$

41. Factor completely: $6x^3y + 3x^2y - 3xy$

Solve.

42. $\dfrac{3x - 2}{x + 2} = \dfrac{8}{3x + 2}$

43. $\dfrac{1}{x} + \dfrac{1}{x - 1} = \dfrac{x^2 - 4}{x^2 - x}$

List the zeros of each function and tell which, if any, are double or triple zeros.

44. $y = -3(x - 2)^3$

45. $y = x(x - 1)(x + 4)$

46. $y = (2x + 1)(x + 3)^2$

Identify which of the functions in Exercises 44–46 are graphed below.

47.

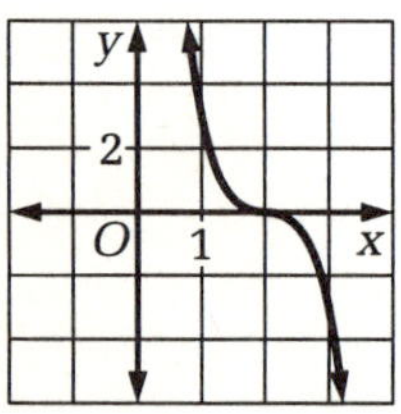

48.

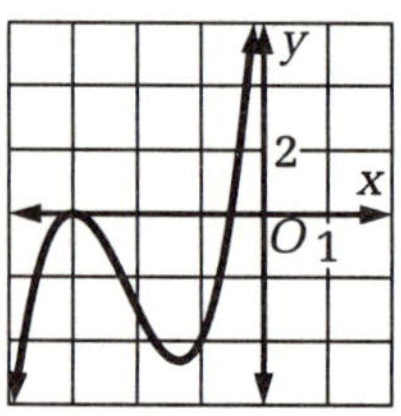

49.

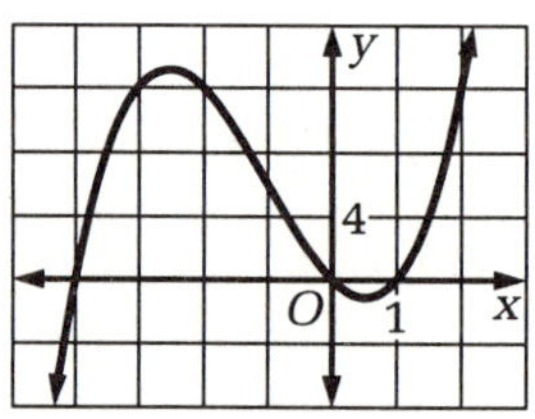

50. Solve $2x^3 + 5x^2 - 12x = 0$.

51. Graph the parametric equations $x = 1 - t$ and $y = t^2 - 1$.

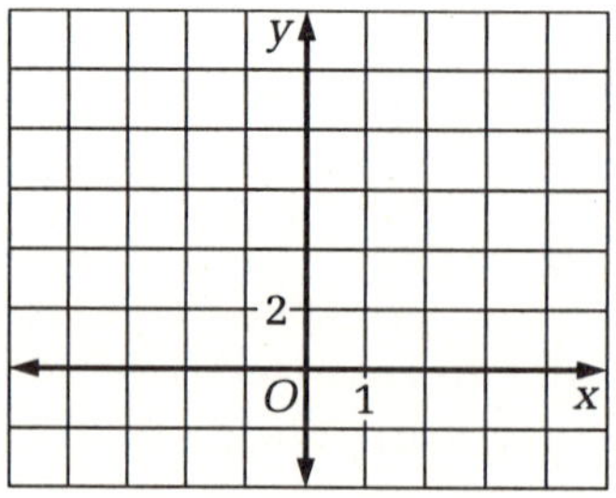

Answers

36. ____________

37. ____________

38. ____________

39. ____________

40. ____________

41. ____________

42. ____________

43. ____________

44. ____________

45. ____________

46. ____________

47. ____________

48. ____________

49. ____________

50. ____________

51. *See question.*

Objectives for *Integrated Mathematics 2*

OBJECTIVES BY UNIT

Unit 1 Sampling and Reasoning

- Recognize a variety of surveying techniques (convenience, random, systematic, stratified, cluster) and know what makes a sample random.
- Ask critical questions necessary for interpreting survey results accurately, such as how was the sample chosen and what was the sample's size.
- Apply inductive reasoning to a variety of real-life and mathematical situations.
- Understand the difference between inductive reasoning and deductive reasoning.
- Recognize errors in logic used in everyday life.

Unit 2 Models of Variation and Growth

- Identify the domain and range of a function.
- Recognize a table, a graph, or an equation as a mathematical model used to describe a real-world situation.
- Understand that inverse variation involves two variables with a constant product and use a data table to determine if two quantities vary inversely.
- Simplify expressions with zero, negative, and fractional exponents.
- Recognize the features of graphs of exponential functions with base 2 and base $\frac{1}{2}$. Recognize exponential growth and decay.

Unit 3 Linear Systems and Matrices

- Interpret linear graphs in real-world contexts (slope, intercepts).
- Solve a linear system by substitution, addition-or-subtraction, or graphing.
- Solve problems involving scalar multiplication, addition, subtraction, or multiplication of matrices.
- Represent a system of equations with a matrix equation.

Unit 4 Quadratic Functions and Graphs

- Graph equations of the form $y = ax^2 + bx + c$, where $a \neq 0$.
- Describe how the graph of $y = ax^2$ relates to the graph of $y = ax^2 \pm k$, $y = a(x \pm h)^2$, or $y = a(x \pm h)^2 \pm k$.
- Solve quadratic equations by factoring.
- Solve problems using the quadratic formula.
- Solve problems involving quadratic systems.

Unit 5 Coordinate Geometry and Quadrilaterals

- Recognize set inclusion relationships among special quadrilaterals.
- Apply the distance formula and slope formula to demonstrate properties about specific geometric figures.
- Use the midpoint formula and properties of perpendicular lines to demonstrate properties about specific geometric figures.
- Examine the effects of translations, rotations, reflections, and dilations on the properties of triangles and quadrilaterals.
- Verify properties of quadrilaterals and triangles using coordinate geometry.

Objectives for *Integrated Mathematics 2* *(continued)*

Unit 6 Counting Strategies, Probability, Binomials

- Solve counting problems by making tree diagrams, tables, and systematic lists.
- Find the probability of events involving the multiplication counting principle, permutations, and combinations.
- Compute probabilities for binomial experiments.
- Find the powers of binomials.

Unit 7 Logic and Proof

- Use conjunctions, disjunctions, and negations.
- Recognize the direct argument, indirect argument, chain rule, and *or* rule as valid arguments, and recognize the following arguments as invalid:

  If p is true, then q is true. q is true. Therefore, p is true.
  If p is true, then q is true. p is not true. Therefore, q is not true.
- Recognize and write good definitions.
- Know postulates of algebra and use them in writing proofs.

Unit 8 Similar and Congruent Triangles

- Read, complete, and write proofs using parallel line theorems and their converses.
- Apply properties of similar triangles in numerical problems.
- Prove that triangles are congruent using ASA, AAS, SSS and SAS.
- Discover and apply properties of 45°-45°-90° and 30°-60°-90° triangles.
- Use the sine, cosine, and tangent functions to find unknown measures in right triangles.

Unit 9 Polynomial and Rational Functions

- Identify terms associated with polynomials: degree, term, standard form of polynomial.
- Solve rational equations that simplify to linear or quadratic equations and recognize an extraneous solution of a rational equation or application.
- Analyze the characteristics of the graph of a cubic function to match the equation of a cubic function with its graph.
- Solve cubic equations by graphing or by factoring.
- Use parametric equations to solve problem situations.

Unit 10 Coordinates and Figures in Space

- Describe the figures formed by rotating plane figures around various axes.
- Describe the sets of points that fit one or more conditions based on distance from one or two points, one or two lines, or one or two planes.
- Use the midpoint and distance formulas in three dimensions.
- Write the equations of circles and of spheres centered at the origin.

Objectives for *Integrated Mathematics 2* *(continued)*

OBJECTIVES BY STRANDS

Algebra

- Identify the domain and range of a function. *(Unit 2)*
- Recognize a table, a graph, or an equation as a mathematical model used to describe a real-world situation. *(Unit 2)*
- Understand that inverse variation involves two variables with a constant product and use a data table to determine if two quantities vary inversely. *(Unit 2)*
- Simplify expressions with zero, negative, and fractional exponents. *(Unit 2)*
- Recognize the features of graphs of exponential functions with base 2 and base $\frac{1}{2}$. Recognize exponential growth and decay. *(Unit 2)*
- Interpret linear graphs in real-world contexts (slope, intercepts). *(Unit 3)*
- Solve a linear system by substitution, addition-or-subtraction, or graphing. *(Unit 3)*
- Solve problems involving scalar multiplication, addition, subtraction, or multiplication of matrices. *(Unit 3)*
- Represent a system of equations with a matrix equation. *(Unit 3)*
- Graph equations of the form $y = ax^2 + bx + c$, where $a \neq 0$. *(Unit 4)*
- Describe how the graph of $y = ax^2$ relates to the graph of $y = ax^2 \pm k$, $y = a(x \pm h)^2$, or $y = a(x \pm h)^2 \pm k$. *(Unit 4)*
- Solve quadratic equations by factoring. *(Unit 4)*
- Solve problems using the quadratic formula. *(Unit 4)*
- Solve problems involving quadratic systems. *(Unit 4)*
- Find the powers of binomials. *(Unit 6)*
- Know postulates of algebra and use them in writing proofs. *(Unit 7)*
- Identify terms associated with polynomials: degree, term, standard form of a polynomial. *(Unit 9)*
- Solve rational equations that simplify to linear or quadratic equations and recognize an extraneous solution of a rational equation or application. *(Unit 9)*
- Analyze the characteristics of the graph of a cubic function to match the equation of a cubic function with its graph. *(Unit 9)*
- Solve cubic equations by graphing or by factoring. *(Unit 9)*
- Use parametric equations to solve problem situations. *(Unit 9)*
- Use the midpoint and distance formulas. *(Units 5 and 10)*

Geometry

- Apply inductive reasoning to a variety of real-world and mathematical situations. (Unit 1)
- Graph equations of the form $y = ax^2 + bx + c$, where $a \neq 0$. *(Unit 4)*
- Describe how the graph of $y = ax^2$ relates to the graph of $y = ax^2 \pm k$, $y = a(x \pm h)^2$, or $y = a(x \pm h)^2 \pm k$. *(Unit 4)*
- Recognize set inclusion relationships among special quadrilaterals. *(Unit 5)*
- Apply the distance formula and slope formula to demonstrate properties about specific geometric figures. *(Unit 5)*

Objectives for *Integrated Mathematics 2* *(continued)*

- Use the midpoint formula and properties of perpendicular lines to demonstrate properties about specific geometric figures. *(Unit 5)*
- Examine the effects of translations, rotations, reflections, and dilations on the properties of triangles and quadrilaterals. *(Unit 5)*
- Verify properties of quadrilaterals and triangles using coordinate geometry. *(Unit 5)*
- Recognize and write good definitions. *(Unit 7)*
- Read, complete, and write proofs using parallel line theorems and their converses. *(Unit 8)*
- Apply properties of similar triangles in numerical problems. *(Unit 8)*
- Prove that triangles are congruent using ASA, AAS, SSS and SAS. *(Unit 8)*
- Discover and apply properties of 45°-45°-90° and 30°-60°-90° triangles. *(Unit 8)*
- Use the sine, cosine, and tangent functions to find unknown measure in right triangles. *(Unit 8)*
- Describe the figures formed by rotating plane figures around various axes. *(Unit 10)*
- Describe the sets of points that fit one or more conditions based on distance from one or two points, one or two lines, or one or two planes. *(Unit 10)*
- Use the midpoint and distance formulas in three dimensions. *(Unit 10)*
- Write equations of circles and of spheres centered at the origin. *(Unit 10)*

Statistics/Probability

- Recognize a variety of surveying techniques (convenience, random, systematic, stratified, cluster) and know what makes a sample random. *(Unit 1)*
- Ask critical questions necessary for interpreting survey results accurately, such as how was the sample chosen and what was the sample's size. *(Unit 1)*
- Find the probability of events involving the multiplication counting principle, permutations, and combinations. *(Unit 6)*
- Compute probabilities for binomial experiments. *(Unit 6)*

Discrete Mathematics

- Apply inductive reasoning to a variety of real-life and mathematical situations. *(Unit 1)*
- Recognize a table, a graph, or an equation as a mathematical model used to describe a real-world situation. *(Unit 2)*
- Solve problems involving scalar multiplication, addition, subtraction, or multiplication of matrices. *(Unit 3)*
- Solve counting problems by making tree diagrams, tables, and systematic lists. *(Unit 6)*
- Find the probability of events involving the multiplication counting principle, permutations, and combinations. *(Unit 6)*
- Compute probabilities for binomial experiments. *(Unit 6)*

Trigonometry

- Discover and apply properties of 45°-45°-90° and 30°-60°-90° triangles. *(Unit 8)*
- Use the sine, cosine, and tangent functions to find unknown measures in right triangles. *(Unit 8)*

Answers

Answers to Inventories questions are referenced to *Integrated Mathematics 2 Study Guide* lessons related to the skill tested by the question.

Toolbox Practice

Skill 1

1.

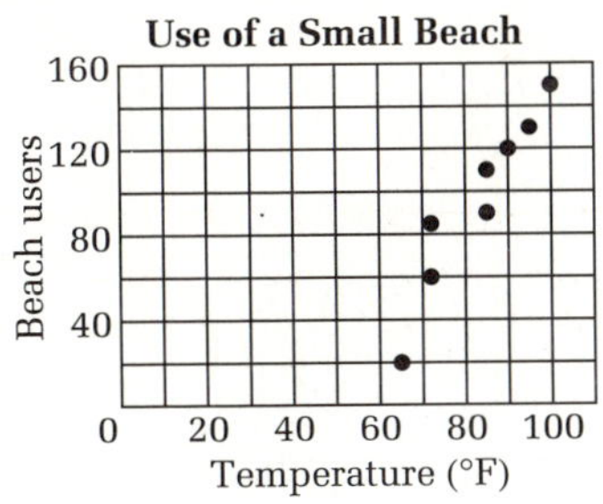

2. positive correlation

3–7. Answers may vary. Examples are given.

3.

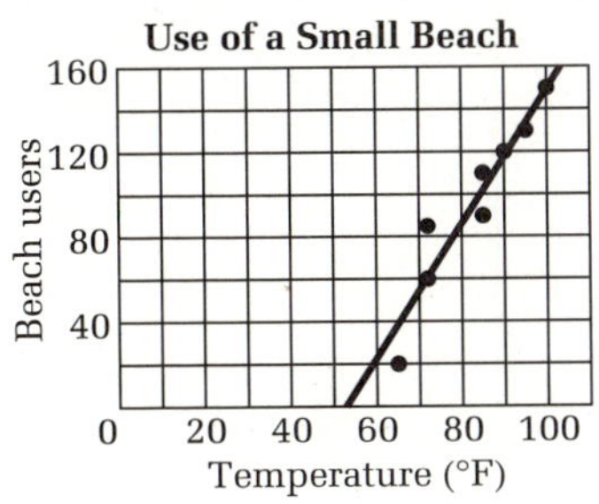

4. about 85 people 5. about 70 people 6. about 160 people 7. about 20 people

Skill 2

1. 30 2. 100 3. 11 4. \$20,000–\$29,000; \$50,000 or over

Skill 3

1. the numbers 70, 71, 72, 75, 76

2. **Finishing Times (minutes)**

Stem	Leaves
3	2 4
4	1 3 4 5 5 6
5	0 1 1 2 2 8
6	2 3 5 6 6
7	1

3. 40–49 and 50–59; 70–79

Skill 4

1. about 22.7 2. 21 3. 10, 18, and 24 4. 32

Skill 5

1.

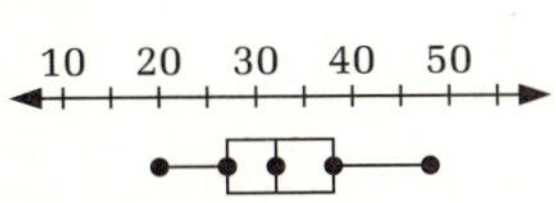

2. 38 3. 27

Skill 6

1. about 38% 2. about 27% 3. about 10%
4. about 12% 5. about 18 sweatshirts

Skill 7

1. 20% 2. 60% 3. 80% 4. 100% 5. about 8%
6. about 15% 7. about 23% 8. about 77% 9. 0
10. 100%

Skill 8

1. 15% 2. 50% 3. about 17% 4. about 67%

Skill 9

1. 12 2. 67 3. −6 4. 23 5. 19 6. 83 7. 0
8. 94 9. −21 10. −4 11. 478 12. −72 13. −24
14. 6 15. −200 16. −396 17. −416 18. −256

Skill 10

1. $3x + 7$ 2. $-2y + 17$ 3. $n - 5m$ 4. $-7k + 23$
5. $3x^2 + 3x$ 6. $-3xy + 3yz$ 7. $-6a^2 + ab$
8. $3k^2 + 3hk$ 9. $7x^3 - 7$

Skill 11

1. $2x^2 + 3x$ 2. $-4x^2 + 20x$ 3. $-14a^2 + 42a$
4. $x^2 - 4x - 45$ 5. $y^2 - 8y + 12$ 6. $-x^2 + 13x - 30$
7. $2x^2 - 13x - 7$ 8. $-3n^2 - 8n + 35$
9. $4x^2 - 22x + 10$ 10. $6x^2 - 41x + 44$
11. $14x^2 - 9x - 8$ 12. $15k^2 - 22k + 8$

Skill 12

1. $3\sqrt{5}$ 2. $10\sqrt{3}$ 3. $2\sqrt{19}$ 4. $10\sqrt{3}$ 5. $60\sqrt{5}$
6. $\sqrt{42}$ 7. $3\sqrt{10}$ 8. $4.2\sqrt{33}$ 9. $24\sqrt{6}$ 10. 6
11. 27 12. $12\sqrt{3}$ 13. $60\sqrt{2}$

Skill 13

1. $x = 72$ 2. $n = 6$ 3. $x = 20$ 4. $x = 25$ 5. $x = 11$
6. $n = 5$ 7. $w = 86$ 8. $x = 3$ 9. $x = -5$ 10. $x = 7$
11. $x = \frac{1}{3}$ 12. $x = 16$ 13. $x = \frac{15}{13}$ 14. $n = -21$
15. $x = \frac{1}{3}$

Skill 14

1. $x < 4$

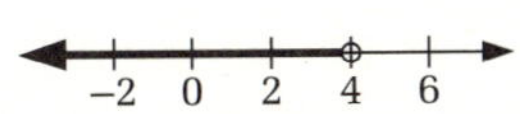

2. $x \geq -2$

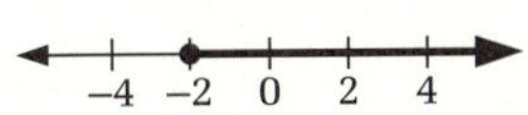

3. $x > -5$ 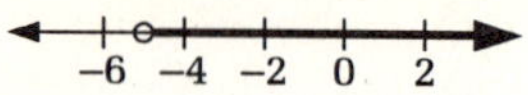

4. $x \geq -2$ 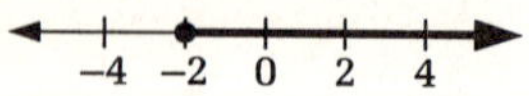

5. $x > -4$ 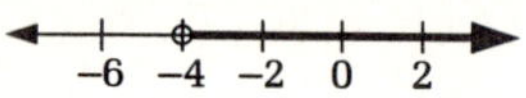

6. $x \geq 1$ 

7. $x < 2$ 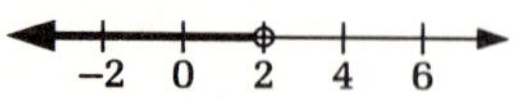

8. $x \leq 0$ 

9. $x < -4$ 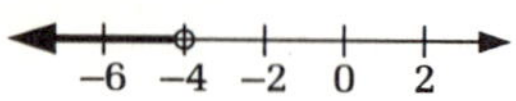

Skill 15

1. $h = \frac{2A}{b}$ **2.** $r = \frac{C}{2\pi}$ **3.** $b = 2A - a$ **4.** $y = 13 - 2x$
5. $y = 4 - \frac{5}{2}x$ **6.** $n = \frac{m}{180} + 2$ **7.** $x = \frac{13}{2} - \frac{y}{2}$
8. $x = \frac{y - b}{m}$

Skill 16

1. $x = \pm 7$ **2.** $x \approx \pm 4.36$ **3.** $x = 2$ **4.** $x = 4$
5. $x = -5$ **6.** $x \approx \pm 3.32$ **7.** $x \approx \pm 9.85$
8. $x \approx \pm 4.90$

Skill 17

1. $x = 15.6$ **2.** $z = 10.8$ **3.** $y = 5.6$ **4.** $x = 11.2$
5. $w = 15.75$ **6.** $y = \frac{40}{11} \approx 3.64$ **7.** $q = 10.2$
8. $n = 52.25$

Skill 18

1. 0, −10 **2.** 0, 4 **3.** 0, 3 **4.** −1, −6 **5.** −4, 7
6. $\frac{1}{2}$, −3

Skill 19

1. Yes. **2.** No. **3.** No. **4.** Yes.

Skill 20

1.

$y = x - 3$		
x	**y**	**(x, y)**
−3	−6	(−3, −6)
−2	−5	(−2, −5)
−1	−4	(−1, −4)
0	−3	(0, −3)
1	−2	(1, −2)
2	−1	(2, −1)
3	0	(3, 0)

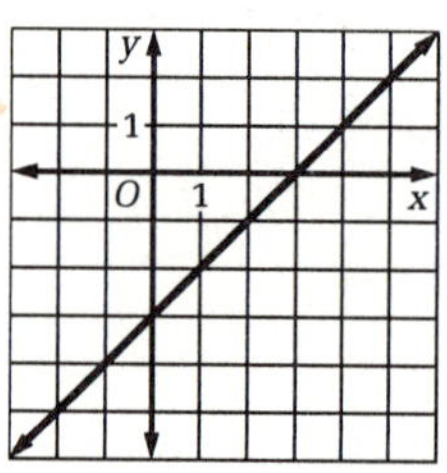

2.

$y = 3x - 4$		
x	**y**	**(x, y)**
−3	−13	(−3, −13)
−2	−10	(−2, −10)
−1	−7	(−1, −7)
0	−4	(0, −4)
1	−1	(1, −1)
2	2	(2, 2)
3	5	(3, 5)

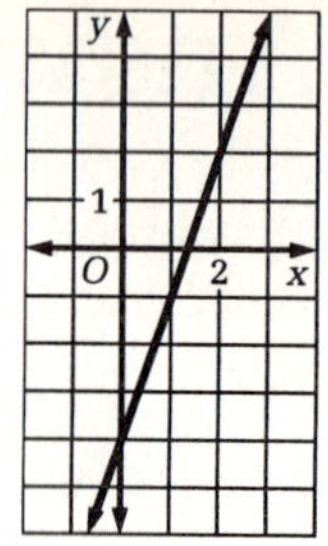

3.

$y = -2x + 3$		
x	**y**	**(x, y)**
−1	5	(−1, 5)
0	3	(0, 3)
1	1	(1, 1)
2	−1	(2, −1)
3	−3	(3, −3)

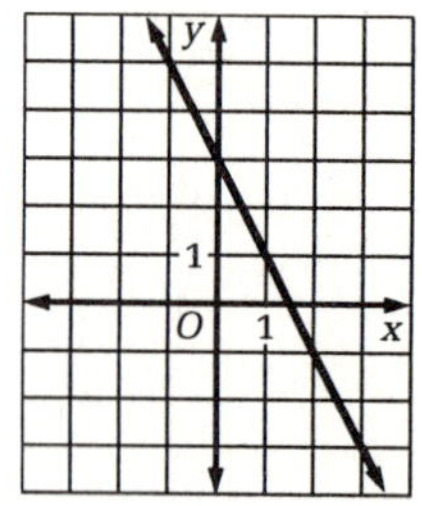

4.

$y = \frac{1}{2}x + 1$		
x	**y**	**(x, y)**
−3	$-\frac{1}{2}$	$\left(-3, -\frac{1}{2}\right)$
−2	0	(−2, 0)
−1	$\frac{1}{2}$	$\left(-1, \frac{1}{2}\right)$
0	1	(0, 1)
1	$1\frac{1}{2}$	$\left(1, 1\frac{1}{2}\right)$
2	2	(2, 2)

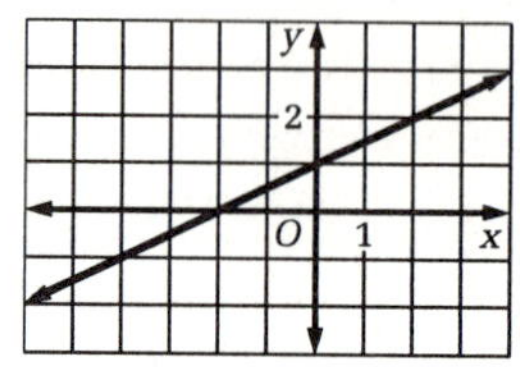

5.

$y = \lvert x \rvert$		
x	**y**	**(x, y)**
−3	3	(−3, 3)
−2	2	(−2, 2)
−1	1	(−1, 1)
0	0	(0, 0)
1	1	(1, 1)
2	2	(2, 2)

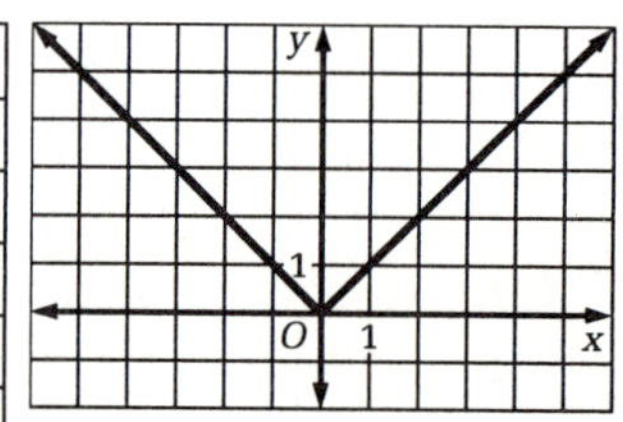

6.

$y = x^2 + 1$		
x	**y**	**(x, y)**
−3	10	(−3, 10)
−2	5	(−2, 5)
−1	2	(−1, 2)
0	1	(0, 1)
1	2	(1, 2)
2	5	(2, 5)
x	y	(x, y)

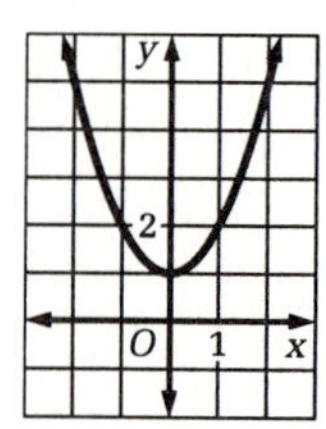

7.

$y = 2x^2 - 5$		
x	**y**	**(x, y)**
–3	13	(–3, 13)
–2	3	(–2, 3)
–1	–3	(–1, –3)
0	–5	(0, –5)
1	–3	(1, –3)
2	3	(2, 3)

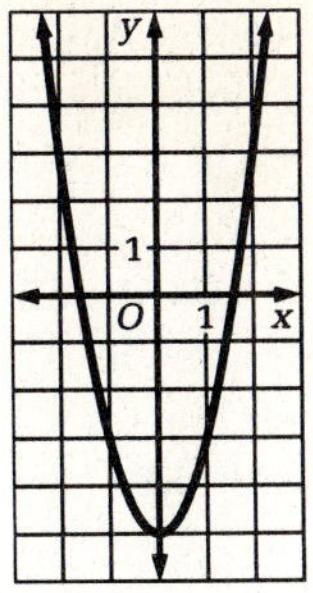

8.

$y = \frac{1}{x}$		
x	**y**	**(x, y)**
–4	$-\frac{1}{4}$	$\left(-4, -\frac{1}{4}\right)$
–3	$-\frac{1}{3}$	$\left(-3, -\frac{1}{3}\right)$
–2	$-\frac{1}{2}$	$\left(-2, -\frac{1}{2}\right)$
–1	–1	(–1, –1)
$-\frac{1}{2}$	–2	$\left(-\frac{1}{2}, -2\right)$
0	—	—
$\frac{1}{2}$	2	$\left(\frac{1}{2}, 2\right)$
1	1	(1, 1)
2	$\frac{1}{2}$	$\left(2, \frac{1}{2}\right)$
3	$\frac{1}{3}$	$\left(3, \frac{1}{3}\right)$
4	$\frac{1}{4}$	$\left(4, \frac{1}{4}\right)$

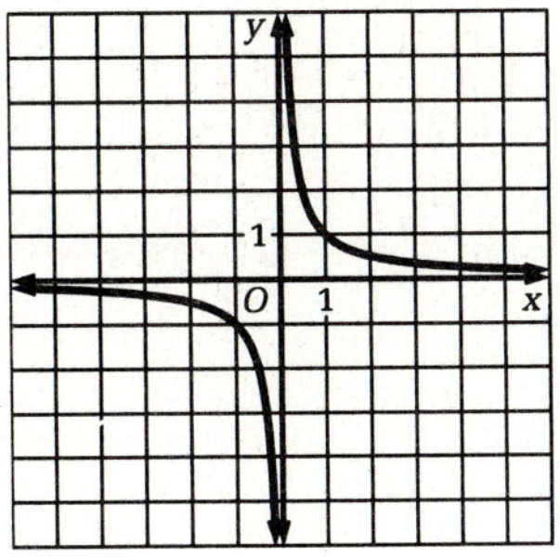

Skill 21

1. $-\frac{2}{3}$ **2.** $\frac{11}{2}$ **3.** $-\frac{2}{5}$ **4.** undefined

Skill 22

1. 161 mi **2.** 4.4 ft

Skill 23

1. slope = 3; vertical intercept = –4
2. slope = $-\frac{5}{2}$ or –2.5; vertical intercept = $\frac{56}{5}$ or 11.2
3. slope = –7; vertical intercept = 6 **4.** slope = 4; vertical intercept = 0 **5.** $y = \frac{1}{2}x + 3$ **6.** $y = -2x + 1$
7. $y = \frac{5}{2}x - 3$

Skill 24

1. vertical intercept = 6; horizontal intercept = 2

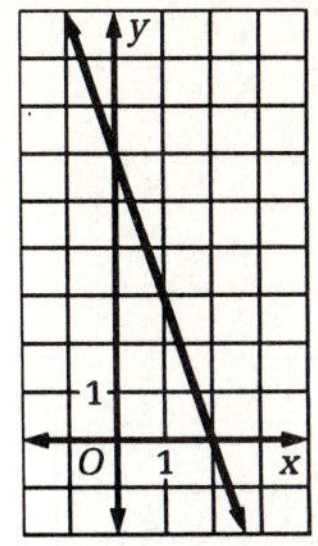

2. vertical intercept = –5; horizontal intercept = 2

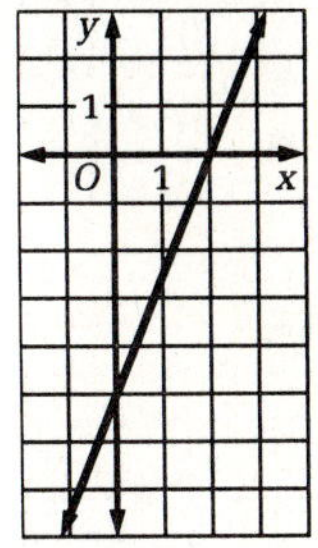

3. vertical intercept = 3; horizontal intercept = –3

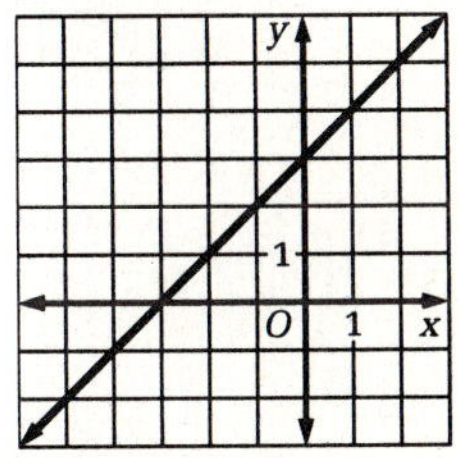

4. vertical intercept = –8; horizontal intercept = 2

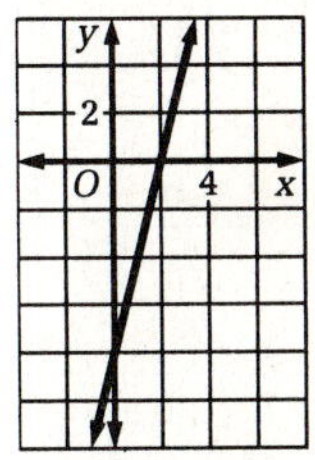

5. vertical intercept = –3; horizontal intercept = $\frac{3}{2}$

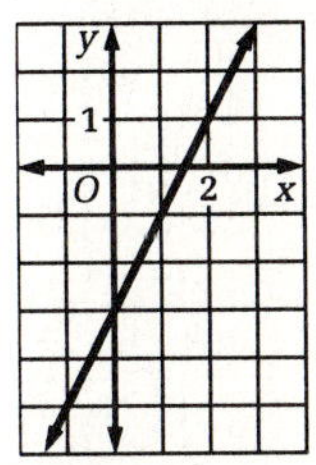

6. vertical intercept = 5; horizontal intercept = 2

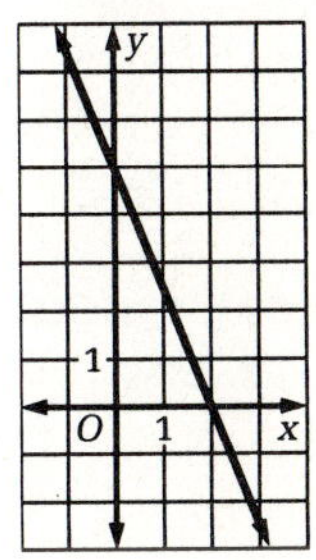

Skill 25

1.

$y = -x + 3$		
x	y	(x, y)
0	3	(0, 3)
3	0	(3, 0)

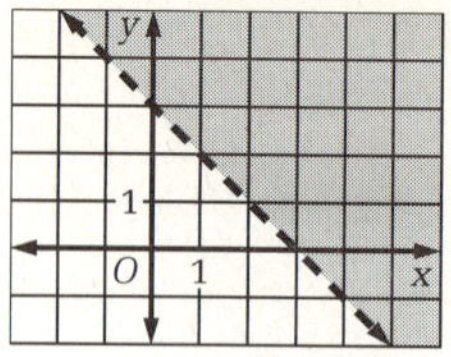

2.

$y = 2x - 4$		
x	y	(x, y)
0	–4	(0, –4)
2	0	(2, 0)

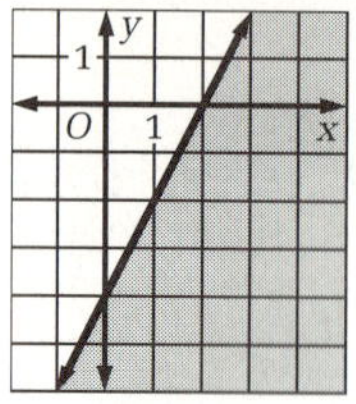

3.

$y = \frac{1}{2}x + 1$		
x	y	(x, y)
0	1	(0, 1)
–2	0	(–2, 0)

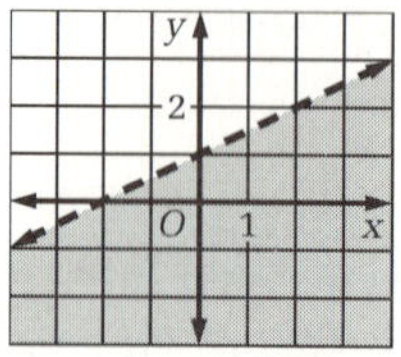

4.

$2x + 3y = 6$		
x	y	(x, y)
0	2	(0, 2)
3	0	(3, 0)

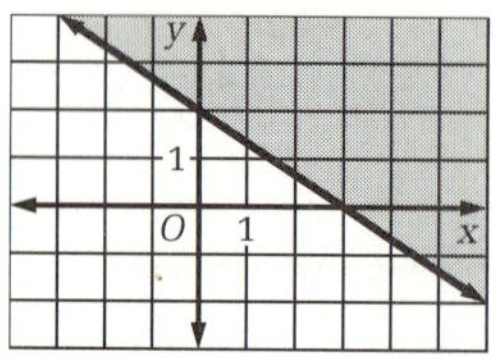

5.

$x - 3y = 3$		
x	y	(x, y)
0	–1	(0, –1)
3	0	(3, 0)

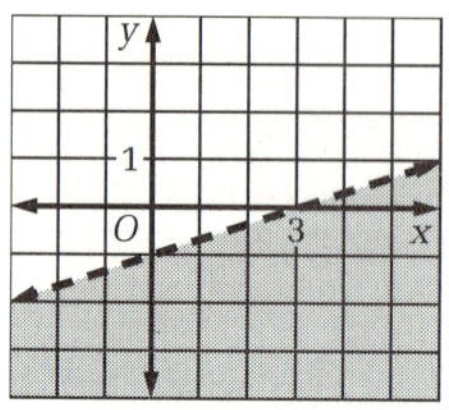

6.

$3x + 2y = 9$		
x	y	(x, y)
0	4.5	(0, 4.5)
3	0	(3, 0)

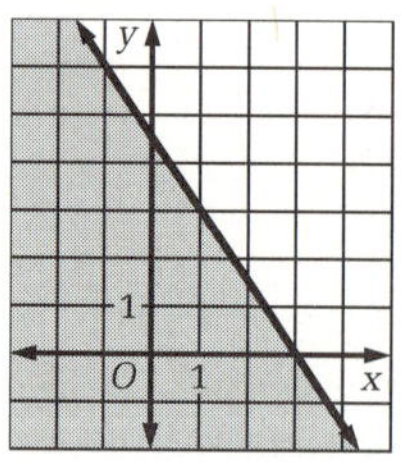

Skill 26

1. 54° **2.** $x° = 64°$, $y°$ 159 116°, $z° = 64°$ **3.** 57°

Skill 27

1. $\angle G$ and $\angle D$ are right angles. Side GF is parallel to side DE. **2.** Side $AB \cong$ side DC. Side $AD \cong$ side BC. **3.** All angles are right angles. Side AD is parallel to side BC. Side DC is parallel to side AB. **4.** Side $AB \cong$ side $BC \cong$ side AC. $\angle A$, $\angle B$, and $\angle C$ are equal in measure. **5.** Side AB is parallel to side DE. $\angle A$ is equal in measure to $\angle CDE$. $\angle B$ is equal in measure to $\angle CED$. $\angle C$ is a right angle. **6.** $\angle ZUV$ is a right angle. Side ZY is parallel to side VW. Side ZU is parallel to side WX. Side $UZ \cong$ side UV.

Skill 28

1. 160 ft^2 **2.** 19.5 $in.^2$ **3.** 9π $in.^2 \approx 28.274$ $in.^2$ **4.** 702 $in.^2$ **5.** 250π $cm^2 \approx 785.4$ cm^2 **6.** 200 cm^3 **7.** 34π $m^3 \approx 106.8$ m^3 **8.** 400 $in.^3$

Skill 29

1. $\frac{4}{8.5} \approx 0.47$ **2.** $\frac{7.5}{8.5} \approx 0.88$ **3.** $\frac{4}{7.5} \approx 0.53$

4. $\frac{7.5}{8.5} \approx 0.88$ **5.** $\frac{4}{8.5} \approx 0.47$ **6.** $\frac{7.5}{4} = 1.875$

7.

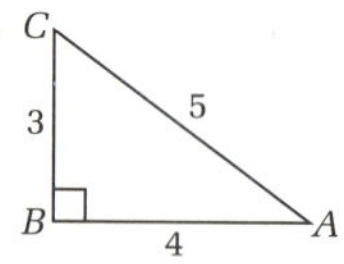

8.

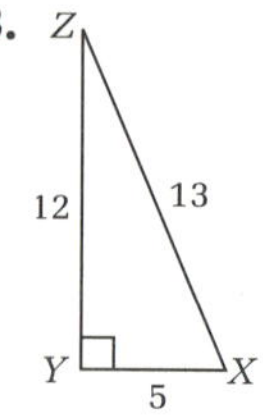

Skill 30

1. $x \approx 26.8$ **2.** $x \approx 5.8$ **3.** $x \approx 26.6$

Skill 31

1. 13 **2.** 10.5 **3.** about 71.1

Skill 32

1. Yes. **2.** Yes. **3.** Yes. **4.** No.

Skill 33

1.

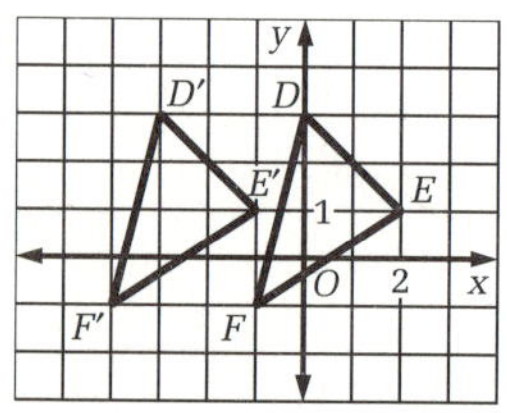

2.

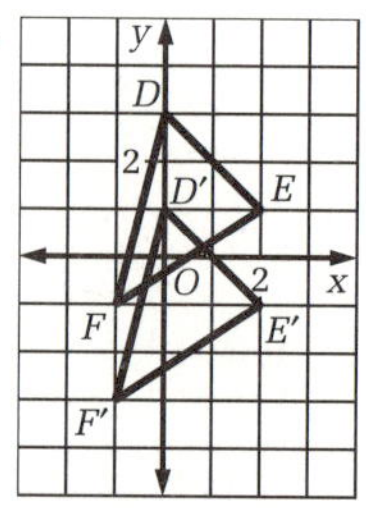

3.

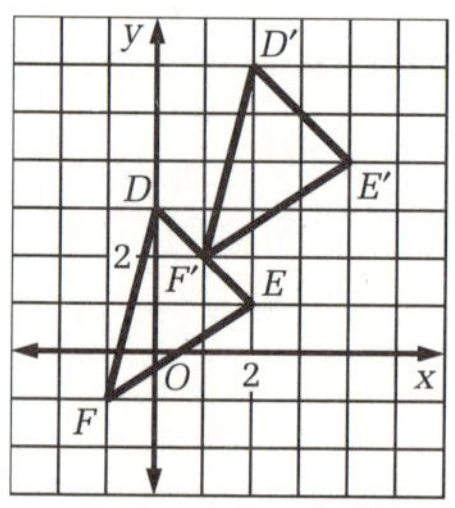

4.

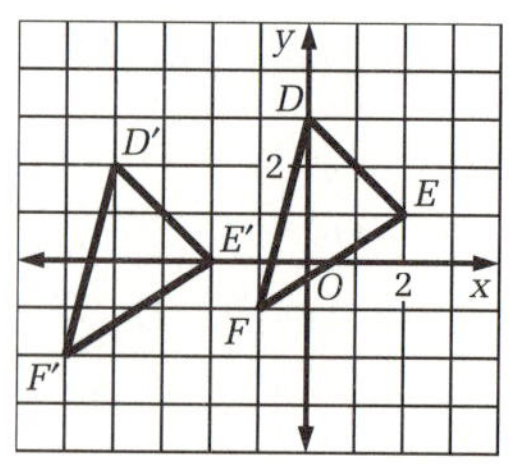

5.

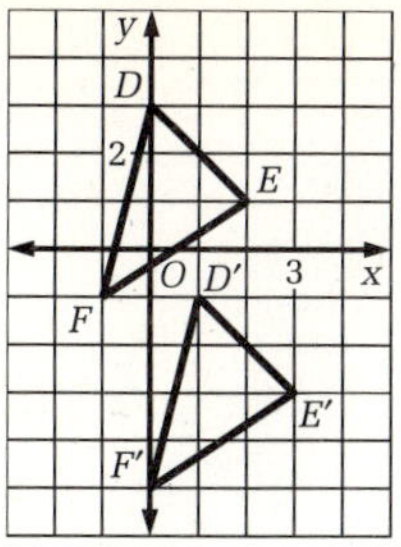

Skill 34

1.

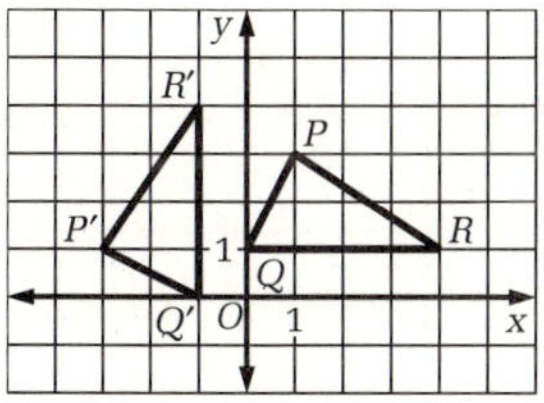

2.

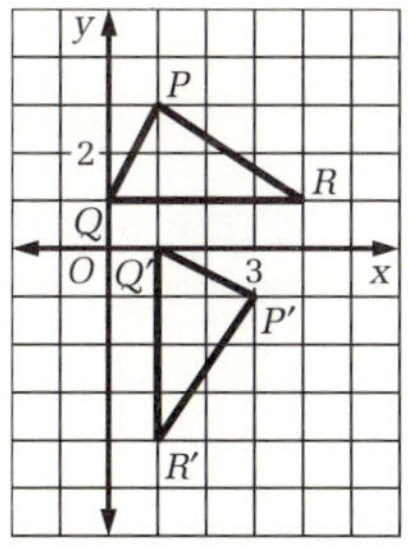

3.

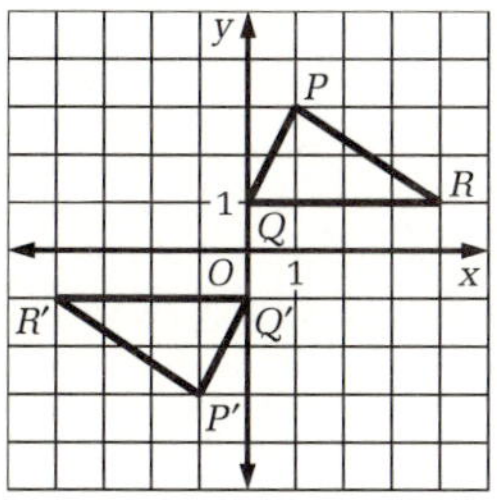

4.

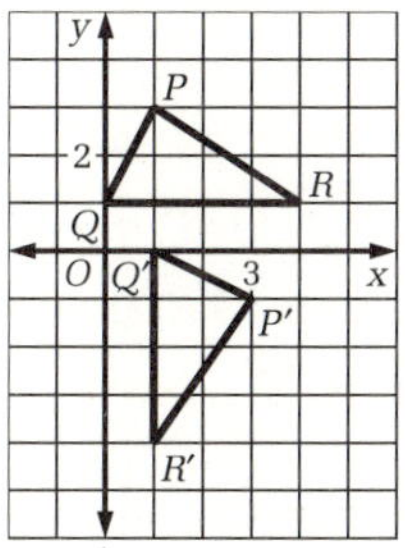

Skill 35

1.

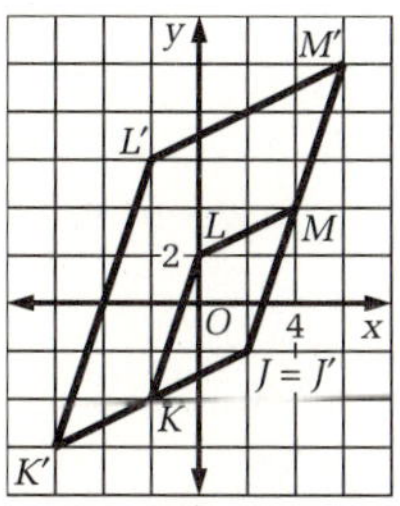

2.

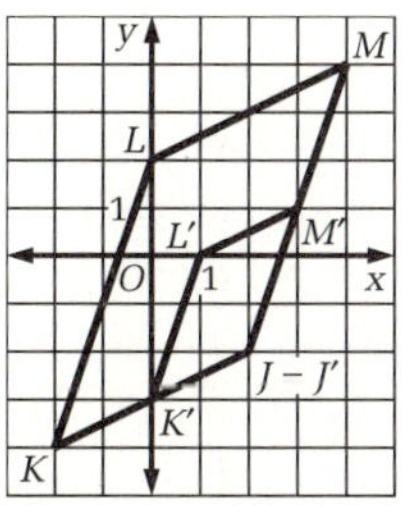

3.

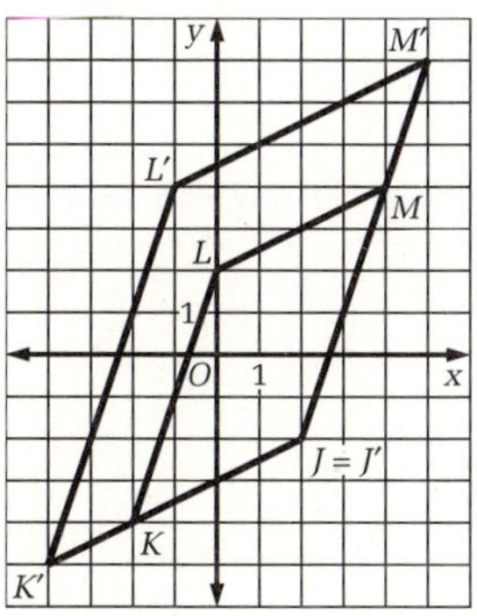

4.

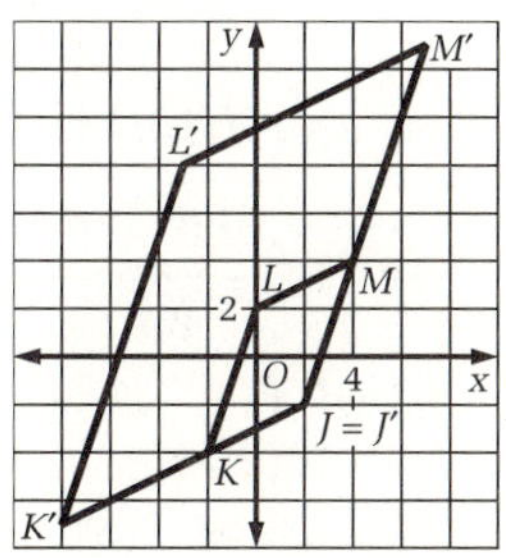

Skill 36

1.

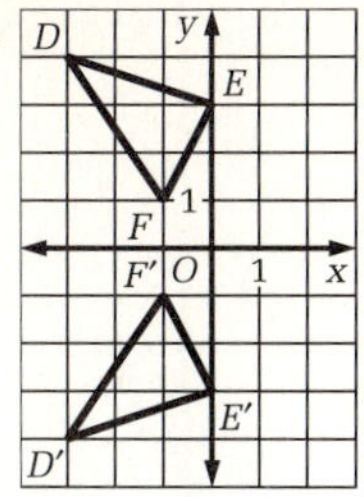

2.

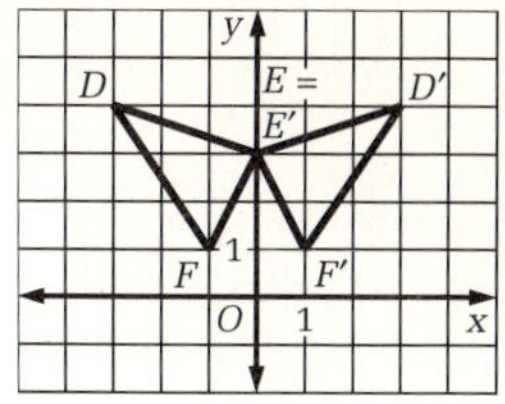

3.

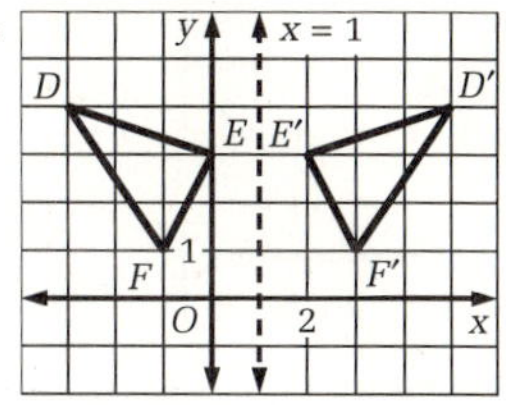

4.

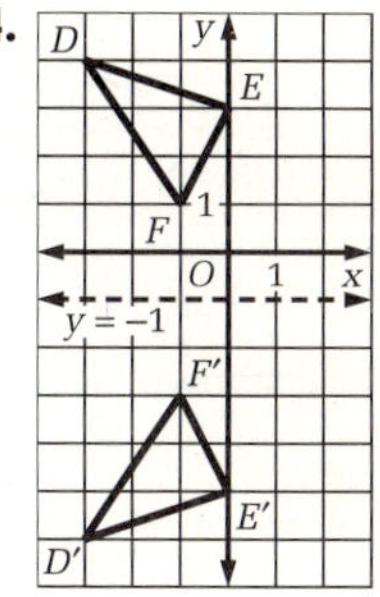

Study Guide Sections

Scatter Plots

1.

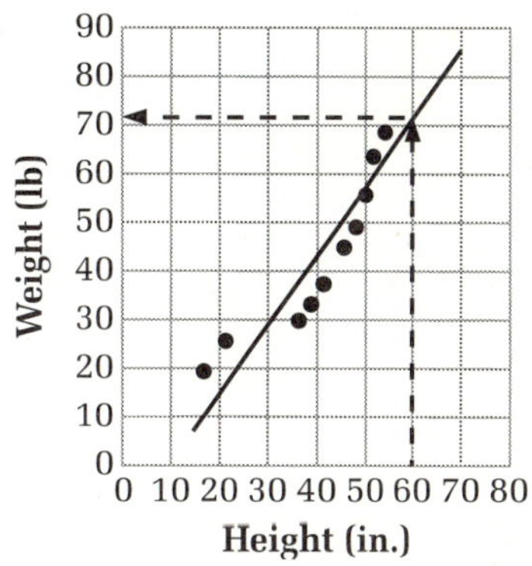

From the graph, the average weight for a girl 60 in. tall is about 71 lb.

2. negative correlation 3. positive correlation

4.

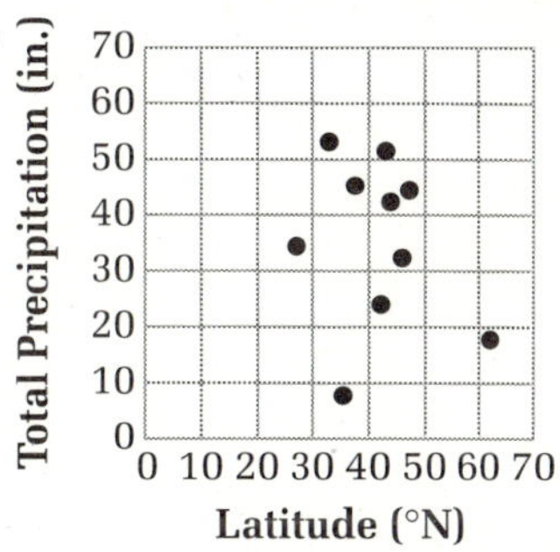

no correlation

5.

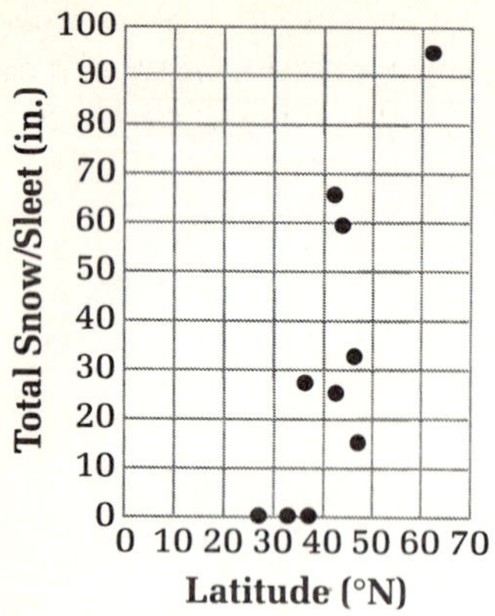

positive correlation

6.

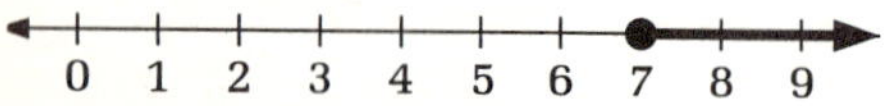

7.

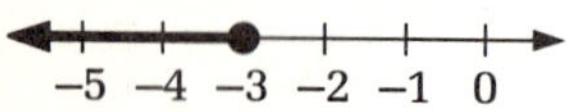

8.

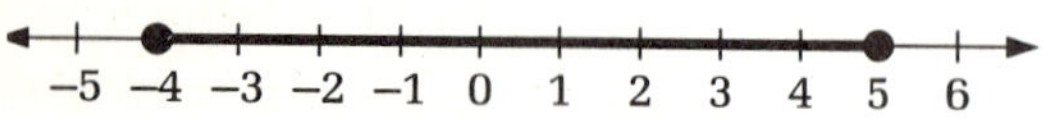

9. a. \$25.50 **b.** \$17.00 **c.** \$4.25y

Histograms and Stem-and-Leaf Plots

1. Book Survey Results

Number of books read	Frequency
0–9	6
10–19	16
20–29	4
30–39	2
40–49	2

2. Book Survey Results

3. Test Scores

5	8
6	0 3 5 5 7 9
7	0 0 1 4 5 5 5 8
8	1 1 3 4 4 7 8
9	0 1 1 3 3 5 6 8

4. Test Scores

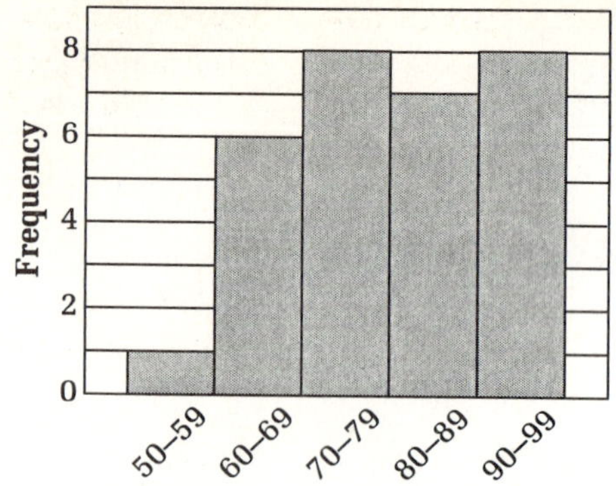

5. $a \geq 16$;

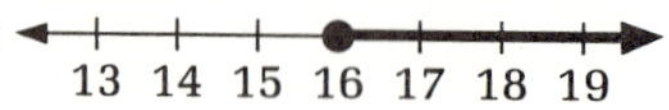

6. $0 < t < 100$;

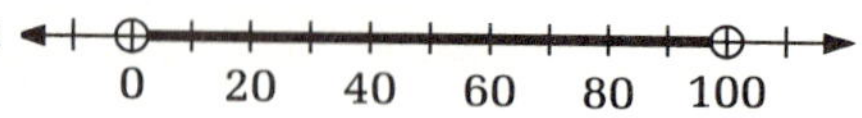

7.

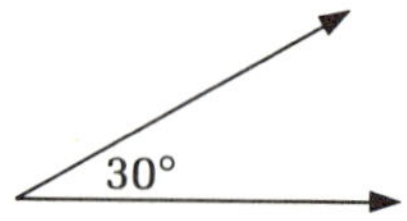

8.

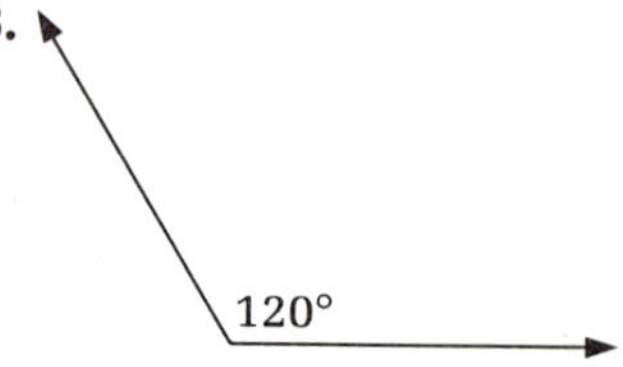

9.

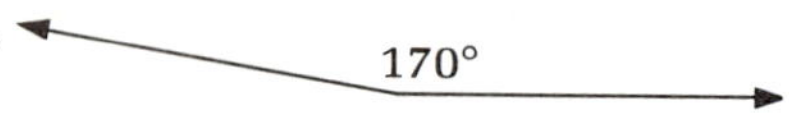

10. median: \$25,000; range: \$49,600; outliers: \$5400 and \$55,000

Mean, Median, and Mode

1. 50 **2.** 76.04 **3.** 76 **4.** 71 and 80 **5.** Yes; 45
6. Answers may vary. A sample answer is given. A: 90–95; B: 81–89; C: 71–80; D: 62–68; F: 45
7. circle graph; Explanations may vary. A sample explanation is given. A circle graph is a good choice for displaying parts (or percents) of a whole. The people who chose each vegetable represent part of the entire group surveyed. **8.** 6^5 **9.** y^4

10.

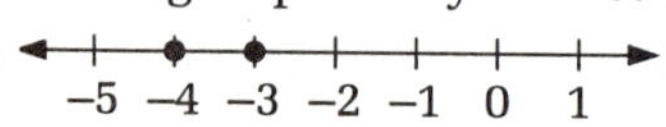

; $-4 < -3$

11.

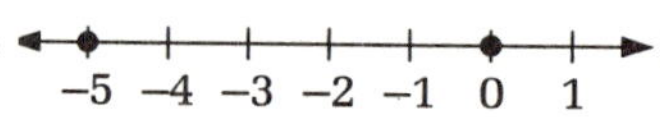

; $-5 < 0$

Box-and-Whisker Plots

1. High Temperatures in Chicago in July

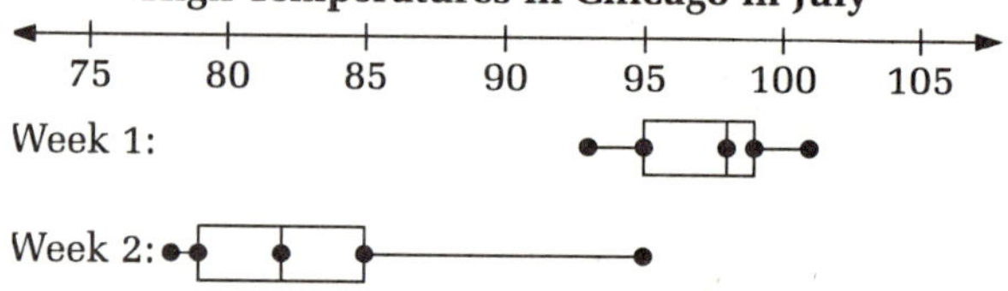

2. The first week has a vary narrow range of high temperatures. The second week has a much greater range of high temperatures that were, except for one very high temperature, much cooler than the first week. **3.** Answers may vary. **4.** 15

5. The \$2000–2499 price range; more notebook computers are priced between \$2000 and \$2499 than the other price intervals. **6.** the actual prices of the 28 notebook computers **7.** $x + y$ **8.** $-2a^2$ **9.** discrete; bar graph

Investigating Probability

1. a. 0.049, or 4.9% **b.** 0.224, or 22.4% **c.** 0
2. about 7 **3.** $\frac{1}{13}$; $\frac{1}{26}$ **4.** $\frac{1}{10}$ **5.** Choice **a** is a pair of complimentary events. **6.** $\frac{3}{5}$ **7.** $\frac{2}{5}$ **8.** $\frac{1}{5}$ **9.** $\frac{4}{5}$
10. Answers may vary. An example is given. Spinning an odd number and spinning an even number. **11.** $\frac{2}{11}$ **12.** $\frac{41}{50}$ **13.** $\frac{3}{8}$ **14.** 6; 3.30
15. 11.18; 5 **16.** 9.54; 4.50 **17.** 3 **18.** 6.8
19. −16

Geometric Probability

1. square A **2.** There are three 30-min time periods between 4:30 P.M. and 6:00 P.M. (4:30–5:00, 5:00–5:30, 5:30–6:00). Therefore, the probability that the call will come while Mita is doing her math homework is $\frac{1}{3}$. **3.** Area of screen: 432 in.^{2}; area of tear: 3 in.^{2}; the probability is $\frac{\text{area of tear}}{\text{area of screen}} = \frac{3}{432} = \frac{1}{144}$. **4.** Segment d is lit for the digits 0, 2, 3, 5, 6, 7, 8, and 9. Therefore, the probability that segment d is lit by a randomly-selected digit is $\frac{8}{10} = \frac{4}{5}$.
5. a. If $12 = x^2$, then $\sqrt{12} = x$. **b.** False; x can also be $-\sqrt{12}$. **6.** Answers may vary. Possible answers are given. The three sides of the triangle are the same length. All three angles have the same measure. The triangle is equilateral. The triangle is isosceles. The triangle is acute. **7.** $A = \frac{1}{2}(b_1 + b_2)h$ **8.** $A = \pi r^2$

Graphs and Functions

1. dependent variable: distance traveled, d; control variable: time, t **2.** dependent variable: distance traveled, d; control variable: rate (speed), r
3. Letting last year be year 1:

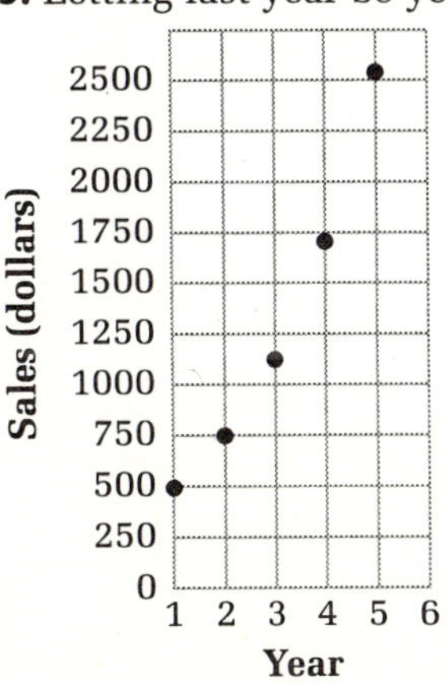

Yes, the graph represents a function since for each year there is only one sales amount.

4.

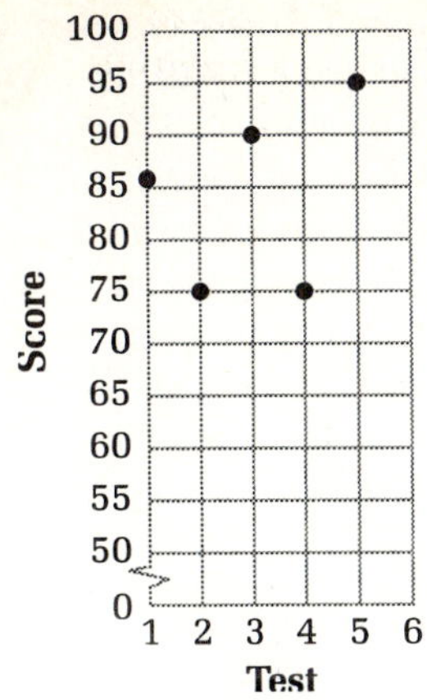

Yes, the graph represents a function since for each test there is only one score.

5. Yes; by the vertical-line test, the graph is a function. **6.** No; by the vertical-line test, the graph is not a function. For example, the vertical line $x = 0$ (the y-axis) intersects the graph in two places.
7. Mark's score is a function of his playing time; dependent variable: s; control variable: playing time, t. **8.** Martha's earnings are a function of the number of the items she assembles; dependent variable: earnings, e; control variable: number of items assembled, n. **9.** Answers may vary. A sample answer is given.

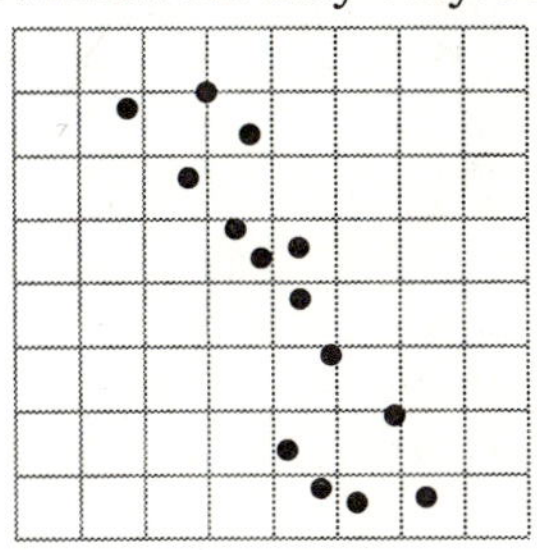

10. $\frac{1}{6}$ **11.** $\frac{4}{3}$ **12.** $-\frac{16}{5}$ **13.** −75, 0, −12
14. 2, −3, −1 **15.** $-\frac{1}{4}$, 1, $\frac{1}{3}$

Exploring Angle Relationships

1. $\overrightarrow{PT}$ and $\overrightarrow{PU}$ **2–5.** Answers may vary. Sample answers are given. **2.** $\angle QPR$ and $\angle UPV$
3. $\angle VPR$, $\angle RPT$, and $\angle SPU$ **4.** $\angle TPU$ and $\angle UPV$
5. $\angle QPS$ and $\angle QPV$ **6.** $x + 78 + 74 = 180$, $x = 28$; 28° **7.** $y + 27 + 90 = 180$, $y = 63$; 63° **8.** $z + 45 + 35 = 180$, $z = 100$; 100° **9.** $a + 87 = 180$, $a = 93$; 93°
10. Ex. 6: acute; Ex. 7: right; Ex. 8: obtuse
11. angle A **12.** is congruent to **13.** ray OP
14. right angle **15.** associative property of addition
16. distributive property **17.** $7x - 35$ **18.** $3x^2 - 5$

Working Together on Congruent Polygons

1. hexagon $PQRSTU \cong$ hexagon $WVUZYX$; side PQ and side WV, side QR and side VU, side RS and side UZ, side ST and side ZY, side TU and side YX, side UP and side XW

2. Answers may vary. Possible answers are given. quadrilateral $BCDE \cong$ quadrilateral $GABF$; side BC and side GA, side CD and side AB, side DE and side BF, side BE and side GF **3.** triangle $WZX \cong$ triangle YZX; side WZ and side YZ, side ZX and side ZX, side WX and side YX **4.** Answers may vary. **5.** $12x - 8$ **6.** $5a^2 - 30a - 35$ **7.** 12 **8.** 43 **9.** $2a + 2b$ or $2(a + b)$ **10.** $2x + 2y$ or $2(x + y)$

Exploring Quadrilaterals and Symmetry

1. rhombus **2.** kite **3.** parallelogram

4.

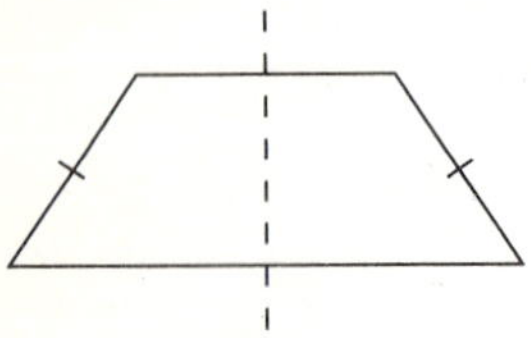

5. no symmetry **6.**

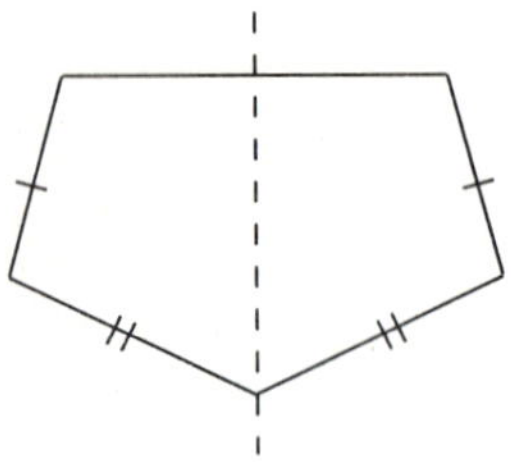

7. Answers may vary. A sample answer is given.

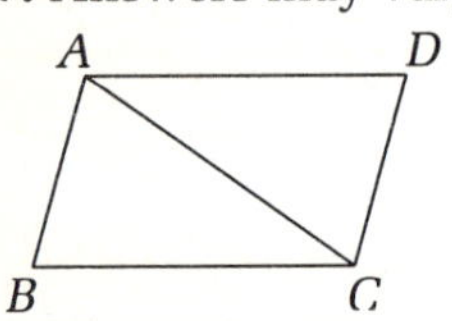

side AB and side CD, side BC and side DA, side AC and side CA

8. $2x + 2y + z$ **9.** $2p + 3q$ **10.** 100,000; 10 to the fifth power **11.** 1,000,000,000,000; 10 to the twelfth power **12.** 100,000,000; 10 to the eighth power **13.** 10; 10 to the first power

Area Formulas

1. 22.5 in.^{2} **2.** 60 ft^2 or $6\frac{2}{3}$ yd^2 **3.** 4.2 m^2 or 42,000 cm^2 **4.** 2 mi **5.** 15 ft **6.** 750 mm^2 **7.** 81 in.^{2} **8.** 3.6 km^2 **9.** 3.2 ft **10.** 64 in.^{2} **11.** 12 m^2 **12.** 216 in.^{2} **13.** sail on left: 94.86 m^2; sail on right: 109.2 m^2; total: 204.06 m^2 **14.** $x = 12$ **15.** $y = -8$ **16.** $x = -36$ **17.** $A'(-1, 2)$; $B'(4, 2)$; $C'(6, -2)$; $D'(-3, -2)$ **18.** $y = \frac{x + 2}{3}$ or $y = \frac{1}{3}x + \frac{2}{3}$ **19.** $y = 3 - x$ **20.** $y = 8x + 32$

Circumference and Arc Length

1. about $1\frac{3}{4}$ ft **2.** 16 ft **3.** 47.1 ft **4.** 4.7 m **5.** 37.7 ft **6.** 20.0 cm **7.** 21 cm **8.** 17.4 ft

9.

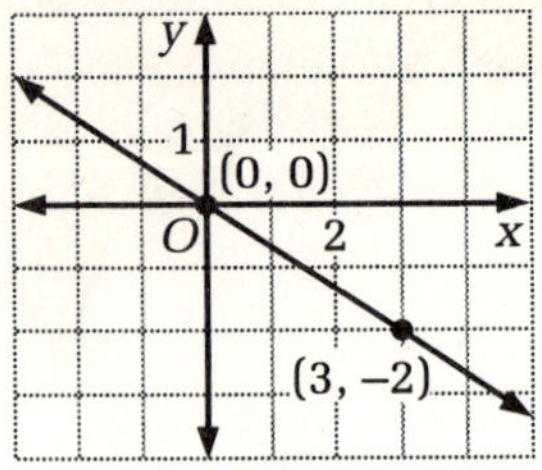

The slope is $-\frac{2}{3}$.

10.

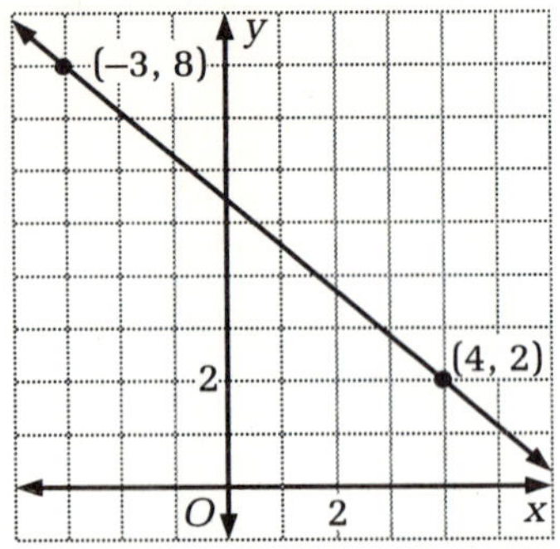

The slope is $-\frac{6}{7}$.

11.

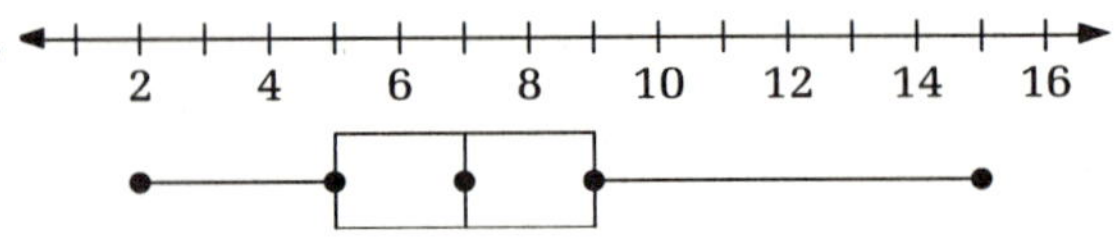

12. 12 **13.** 20

Areas of Circles and Sectors

1. 113.0 in.^{2} **2.** 8.0 km **3.** 4.8 m

4.

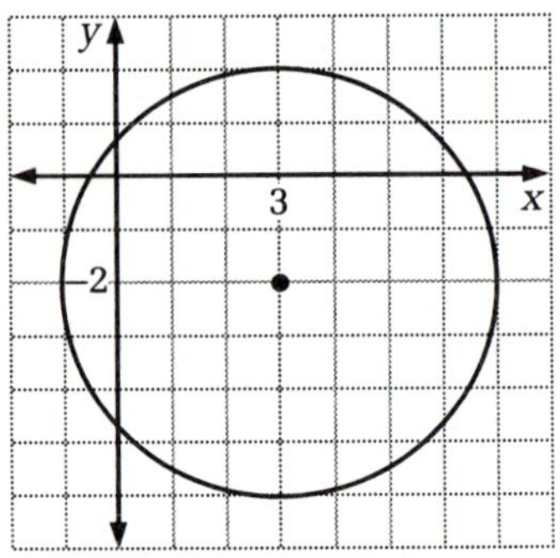

5. 322.1 in.^{2} **6.** 26.2 m^2 **7.** 52.8 km^2 **8.** The slice from the larger pizza is about 0.4 in.^{2} larger. **9.** 0.088 ft/sec **10.** \$1.80/yd **11.** 180°, either clockwise or counterclockwise **12.** –40 **13.** 540 **14.** –7

Surface Area of Space Figures

1. square prism; 54 in.^{2} **2.** triangular prism; 156 cm^2 **3.** about 452 cm^2 **4.** about 52 yd^2 **5.** about 13 cm^2 **6.** 45 ft^2 **7.** about 43 m^2 **8.** $\frac{3}{4}$ **9.** $\frac{1}{4}$ **10.** 1

11.

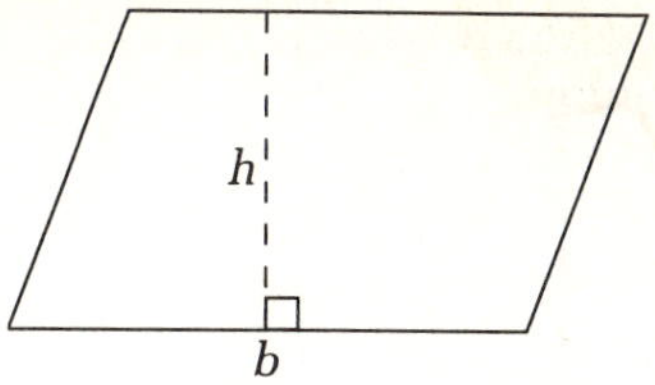

To find the area of the parallelogram, you need to know the length of a side and the distance between that side and the opposite side.

12. 840 in.^{3} **13.** 180 cm^3

Volumes of Prisms and Cylinders

1. 84 cm^3 **2.** 96 ft^3 **3.** 11 cm **4.** 6 yd **5.** about 18.5 m^3 **6.** about 954.3 in.^{3} **7.** about 5.0 mm **8.** about 16.0 ft **9.** rectangular prism; 174 in.^{2} **10.** regular square pyramid; 18 ft^2 **11.** 88 ft/s **12.** $s \approx 12.8$ ft **13.** $y = 2x + 7$ or $2x - y = -7$

Volumes of Pyramids and Cones

1. about 9.23 cm^3 **2.** 50 in.^{3} **3.** about 0.4 m^3 **4.** about 1.77 ft^3 **5.** triangular prism; 20 yd^3 **6.** cylinder; about 15,708 cm^3 **7.** $\frac{15}{6} = 2.5$ **8.** 3 **9.** 12.5 **10.** $x = \pm\frac{8}{3}$

Sine and Cosine Ratios

1. 0.59 **2.** 0.97 **3.** 0.26 **4.** 0.26 **5.** The triangles have two pairs of corresponding angles congruent, so they are similar. **6.** $MN = 7.5$; $PQ \approx 8.7$ **7.** $\sin L = \sin P = \frac{1}{2}$ (or 0.5) **8.** $\cos L = \cos P = \frac{13}{15}$ (or $0.8\overline{6}$) **9.** In similar triangles, the ratios of corresponding lengths are equal. **10.** $\sin 30° = 0.5$; $\cos 30° \approx 0.8660$; They are the same. **11.** about 17.7 yd **12.** $\sin 63° = \frac{22.4}{x}$; 25.1 **13.** $\sin 40° = \frac{x}{95}$; 61.1 **14.** 14.4 in. by 19.2 in.

15.

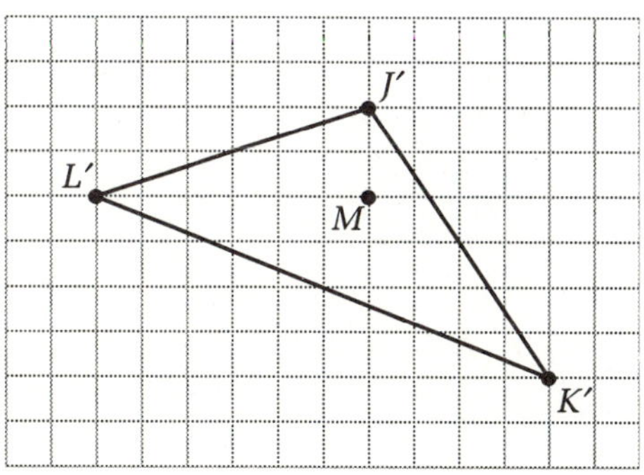

16. Answers may vary. A sample answer is given. A gap in the scale can imply that the bars or lines begin at a value different from the actual value.

17. $\frac{5}{12}$

Direct Variation, Slope, and Tangent

1. 1250 ft **2.** $\frac{4}{3} \approx 1.3$ **3.** 15 ft **4.** $\frac{8}{75} \approx 0.1$ **5. a.** 0.75 **b.** 37° **6. a.** 1 **b.** 45° **7. a.** 0.27 **b.** 15° **8.** 6.7 **9.** 6.3 **10.** $\sin Q = 0.34$; $\cos Q = 0.94$ **11.** 2.6 **12.** 65 **13.** $\frac{5}{13}$ **14.** Let c = child's height at age 2 and a = child's height as an adult; $c = \frac{1}{2}a$.

Translations

1.

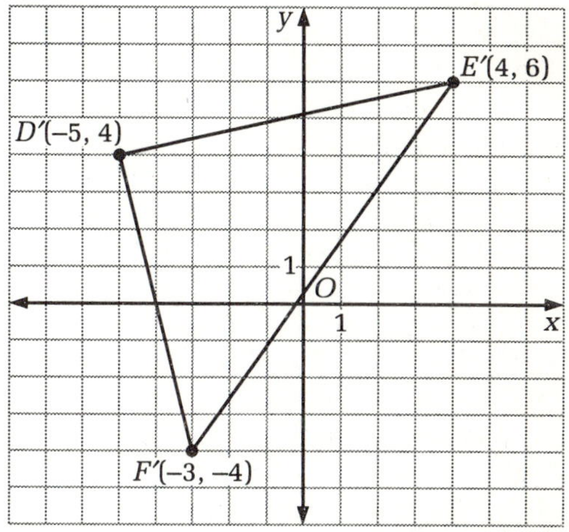

2.

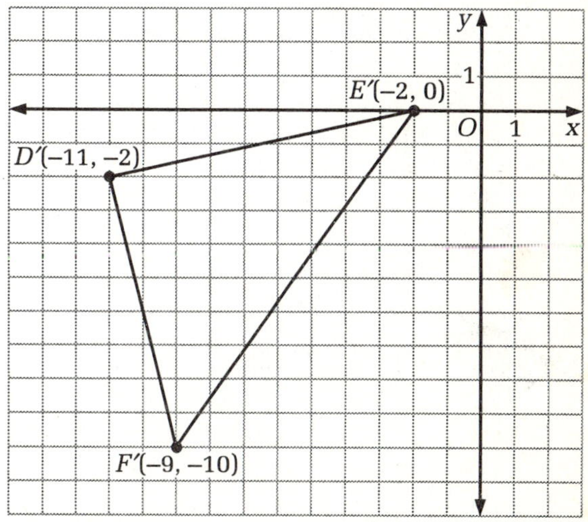

3. Yes; the vertices each move 6 units left. **4.** Yes; the vertices each move 6 units right and 3 units down. **5.** $(x, y) \rightarrow (x - 6, y + 4)$ **6.** $(x, y) \rightarrow (x + 2, y - 7)$ **7.** $P'(2, 1)$ **8.** $P'(-5, 5)$ **9.** $P'(-13, 20)$ **10.** $P'(-3, 11)$

11. a.

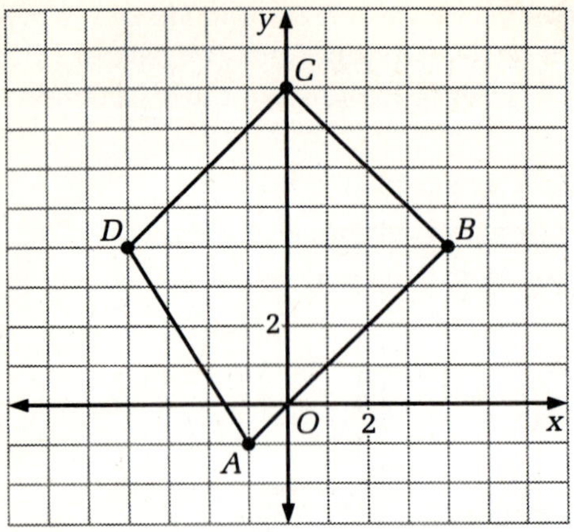

b. Trapezoid; $\overline{AB}$ and $\overline{CD}$ are parallel but $\overline{BC}$ and $\overline{AD}$ are not, so the figure is a quadrilateral with one pair of parallel sides, which is a trapezoid. **12.** $x = 7$ **13.** $a = -10$

Rotations

1. 90° counterclockwise around the origin

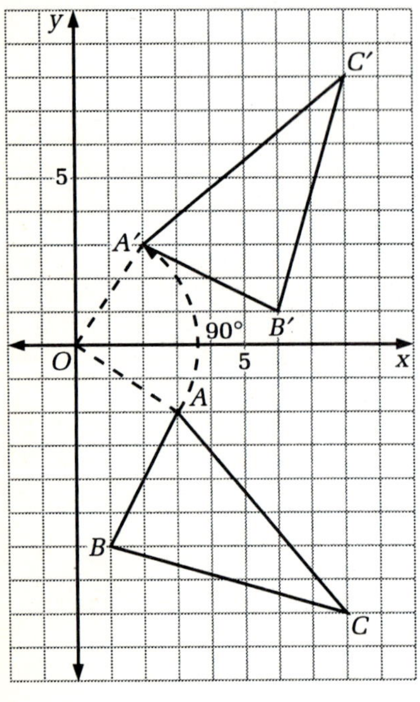

2. 90° clockwise around the origin

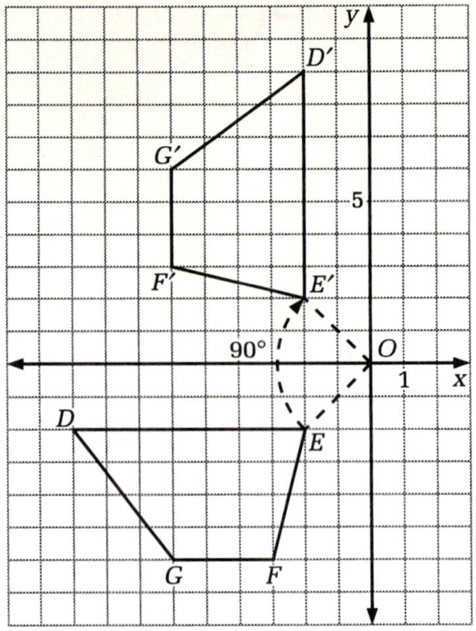

3.

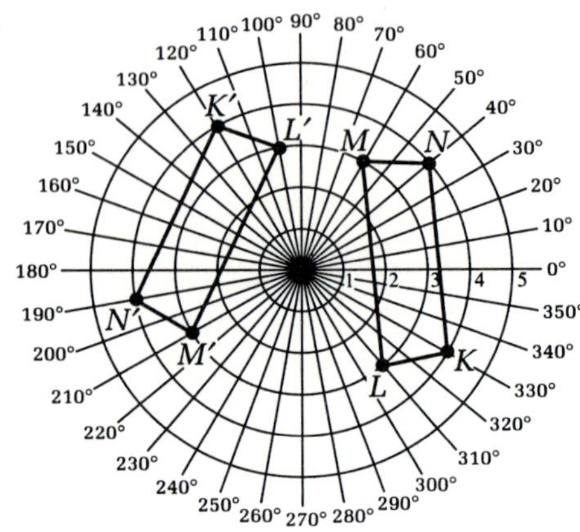

4.

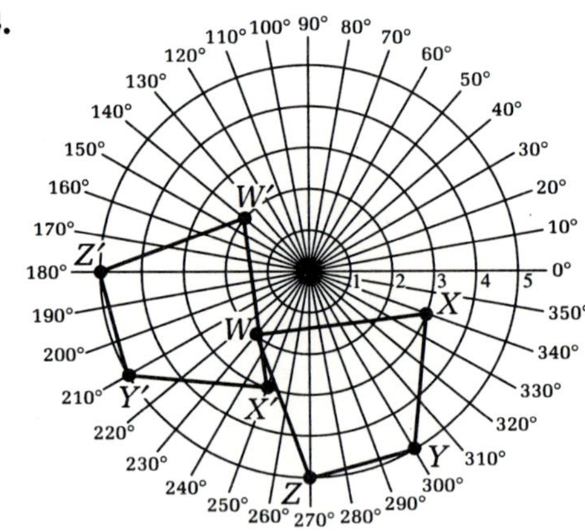

5. Yes. The figure has 180° rotational symmetry.
6. Yes. The figure has 180° rotational symmetry.
7. Yes. The figure has 120° and 240° rotational symmetry.

8.

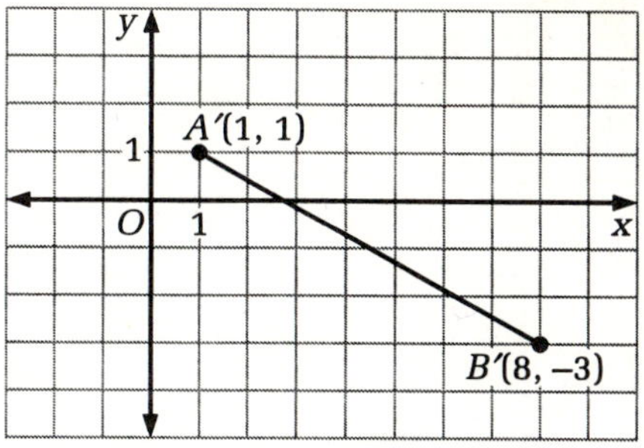

9. Graph may vary. A sample bar graph is given.

National League Baseball, Team ERA (July 10, 1993)

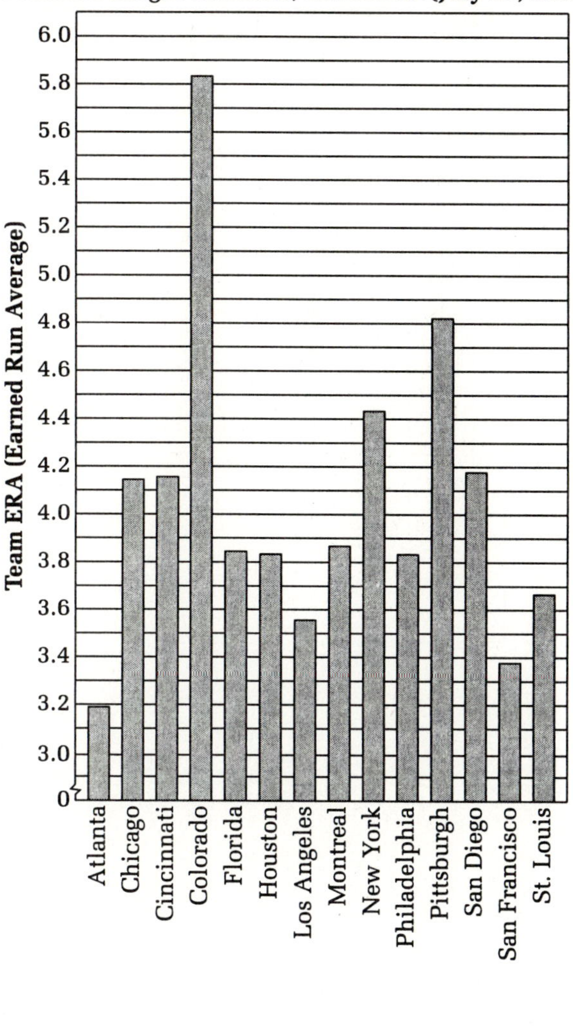

Dilations

1. $\angle A = \angle A' = 116°$; $\angle B = \angle B' = 53°$; $\angle C = \angle C' = 125°$; $\angle D = \angle D' = 66°$; yes

2.

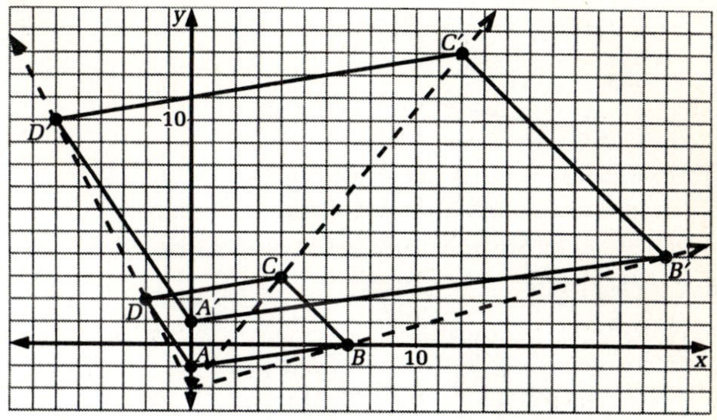

3. yes; yes **4.**

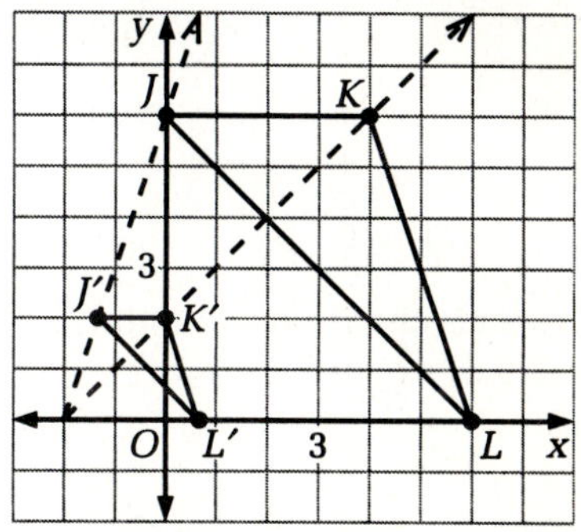

5.

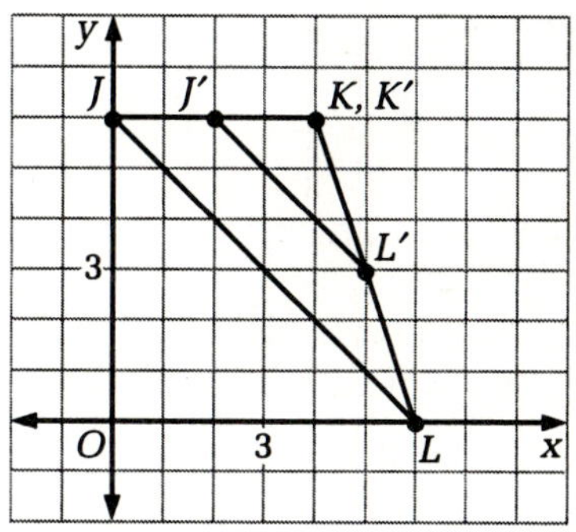

6. yes; no **7.** (0, 2); 2 **8.** (0, 0); 1.5 **9.** $\angle S = 80°$, $\angle T = 40°$, $\angle X = 60°$, $\angle Y = 80°$, $RS = 10.5$, $YZ = 8.1$
10. −4 **11.** 12 **12.** 3 **13.** −4 **14.** $\frac{1}{5}$ **15.** $\frac{1}{4}$ **16.** $\frac{2}{7}$

Reflections

1. reflection **2.** Not a reflection; the orientation of the image is not reversed. **3.** Not a reflection; the triangles are not congruent. **4.** Not a reflection; the dashed line is not a line of symmetry.

5. 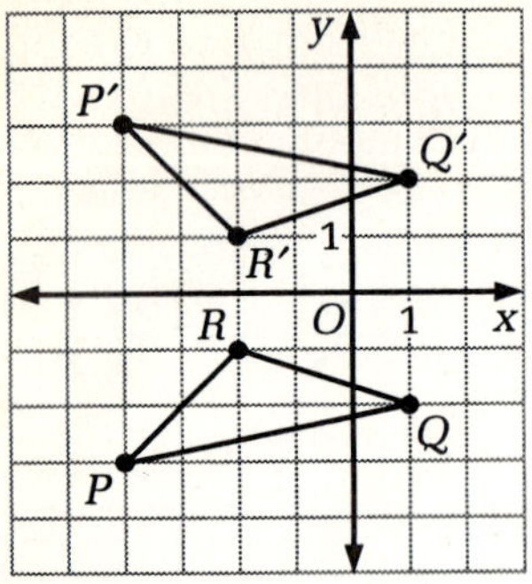

△PQR has a counterclockwise orientation; △$P'Q'R'$ has a clockwise orientation.

6. 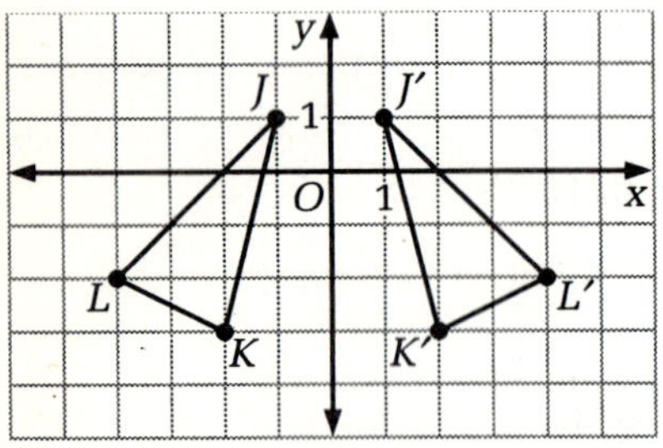

△JKL has a clockwise orientation; △$J'K'L'$ has a counterclockwise orientation.

7. 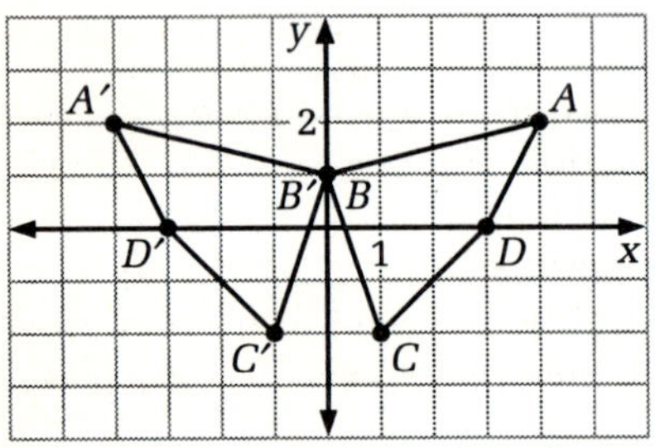

Quadrilateral $ABCD$ has a counterclockwise orientation; quadrilateral $A'B'C'D'$ has a clockwise orientation.

8. 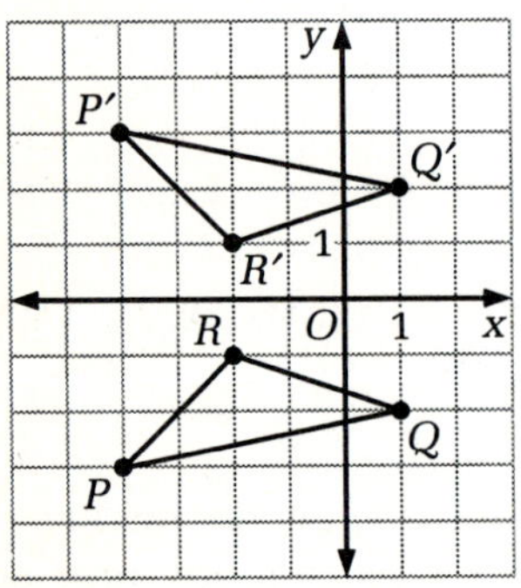

Quadrilateral $OPQR$ has a clockwise orientation; quadrilateral $O'P'Q'R'$ has a counterclockwise orientation.

9. $\frac{64}{729}$ **10.** 48π cm^3

11. 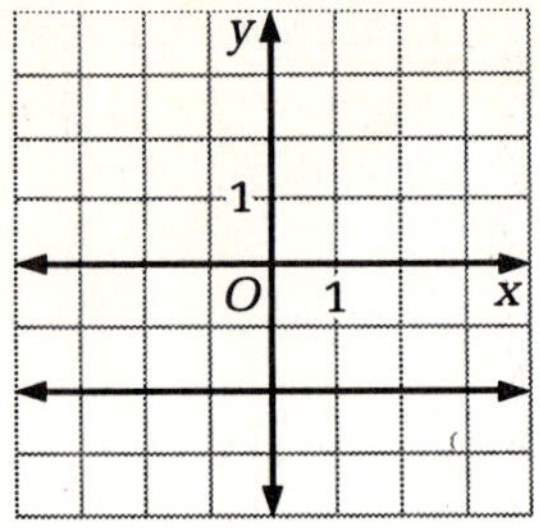

12. 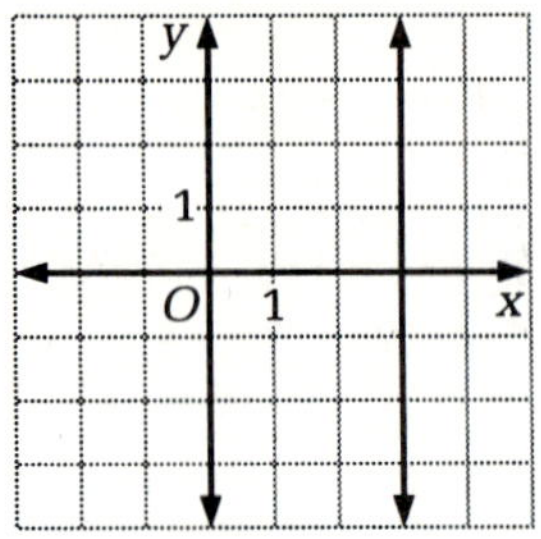

13. 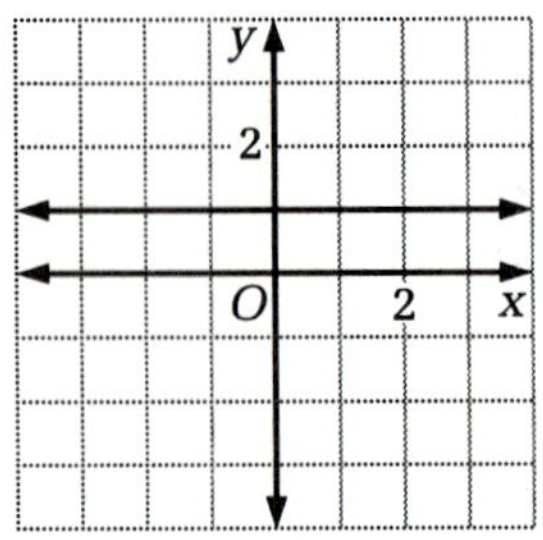

14. 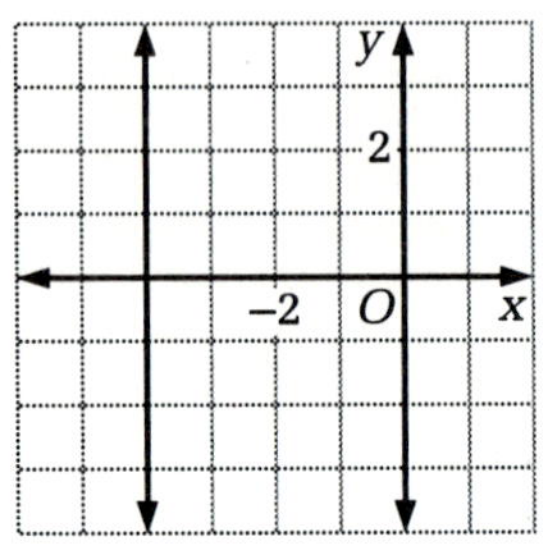

Inventories

Answers to Inventories questions are referenced to *Integrated Mathematics 2 Study Guide* lessons related to the skill tested by the question.

Inventory 1

1. 600 *(SG 1-1, p. 1)* **2.** A *(SG 1-2, p. 4)* **3.** convenience *(SG 1-3, p. 8)* **4.** Answers may vary. An example is given. This sampling method is not at all random, since only people who are shopping on that particular day at that particular time and who feel they are able to spare a few minutes and are willing to talk to the researchers will participate. In addition, the researchers are not likely to ask adults who are accompanied by young children. *(SG 1-4, p. 11)* **5.** 25 *(SG 1-5, p. 14)* **6.** The sum of the first n odd integers is equal to n^2. *(SG 1-5, p. 14)* **7.** inductive *(SG 1-5, p. 14)* **8.** If a figure has four right angles, then it is a square. *(SG 1-6, p. 18)* **9.** False. Counterexamples may vary. Any rectangle that has unequal length and width is a counterexample. *(SG 1-5, p. 14)* **10.** increasing, nonlinear *(SG 2-1, p. 26)* **11.** constant, linear *(SG 2-1, p. 26)* **12.** increasing, linear *(SG 2-1, p. 26)* **13.** decreasing, nonlinear *(SG 2-1, p. 26)* **14.** $y = -x - 1$ *(SG 2-2, p. 30)* **15.** Yes; the product xy is approximately equal to 2 for each ordered pair. *(SG 2-3, p. 35)* **16.** 3942 cm^3 *(SG 2-4, p. 38)* **17.** $h = kt^2$ *(SG 2-5 p. 41)* **18.** $f = \frac{k}{l}$ *(SG 2-3, p. 35)* **19.** $D = kr$ *(SG 2-2, p. 30)* **20.** $\frac{9}{x^5}$ *(SG 2-6, p. 45)* **21.** x^2 *(SG 2-6, p. 45)* **22.** 7 *(SG 2-6, p. 45)* **23.** $-3x^{1/3}$ *(SG 2-6, p. 45)* **24.** $\approx$0.029 g *(SG 2-7, p. 47)*

Inventory 2

1. $\approx$7911 *(SG 1-1, p. 1)* **2.** Answers may vary. An example is given. You could roll the die 25 times and record "the seed germinated" if the die comes up a 1, 2, 3, 4 or 5, and "the seed did not germinate" if the die comes up a 6. *(SG 1-2, p. 4)* **3.** systematic *(SG 1-3, p. 8)* **4.** convenience *(SG 1-3, p. 8)* **5.** stratified random *(SG 1-3, p. 8)* **6.** Answers may vary. An example is given. $n^2 + n + 11$ is a prime number for every positive integer. *(SG 1-5, p. 14)* **7.** False; counterexamples may vary. An example is given. $11^2 + 11 + 11 = 11(13)$, so it is not prime. *(SG 1-5, p. 14)* **8.** If a number is divisible by 2, then it is divisible by 6. *(SG 1-6, p. 17)* **9.** $y = -\frac{2}{3}x + \frac{4}{3}$ *(SG 2-2, p. 30)* **10.** $\approx$11.5 cm^2 *(SG 2-4, p. 38)* **11.** $V = kr^3$ (SG 2-5, p. 41) **12.** $lw = k$ or $l = \frac{k}{w}$ *(SG 2-3, p. 35)* **13.** $s = kr^2$ *(SG 2-5, p. 41)* **14.** $\frac{-25}{2x^2}$ *(SG 2-6, p. 45)* **15.** 4 *(SG 2-6, p. 45)* **16.** $3ab^3$ *(SG 2-6, p. 45)* **17.** $2|r|t^2$ *(SG 2-6, p. 45)* **18.** $y < x$; $y \geq 2$ *(SG 3-1, p. 52)* **19.** no solution *(SG 3-3, p. 59)* **20.** (−6, −13) *(SG 3-2, p. 56)* **21.** (3, −1) *(SG 3-4, p. 63)*

22. $\begin{bmatrix} 0 & 7 & -7 \\ 4 & -9 & -2 \end{bmatrix}$ *(SG 3-5, p. 66)*

23. $\begin{bmatrix} -2 & 6 \\ 0 & -4 \\ -3 & 18 \end{bmatrix}$ *(SG 3-7, p. 75)*

24. dilation with center at the origin by scale factor 0.5 *(SG 3-6, p. 70)* **25.** translation 2 units left and one unit up *(SG 3-6, p. 70)*

Inventory 3

1. $\approx$39 *(SG 1-1, p. 1)*

2. 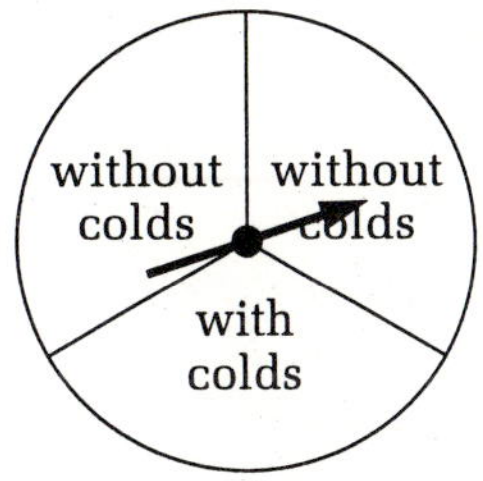

(SG 1-2, p. 4)

3. convenience *(SG 1-3, p. 8)* **4.** Only people who listened to the station and had access to a phone at the given time would hear of the survey and be able to respond. Furthermore, the callers would be people with strong opinions; people who were undecided or did not have strong opinions either way would not call in. *(SG 1-4, p. 11)* **5.** $\frac{8 \cdot 9}{2} = 36$ *(SG 1-5, p. 14)* **6.** $1 + 2 + 3 + \ldots + n = \frac{n(n+1)}{2}$ *(SG 1-5, p. 14)* **7.** inductive *(SG 1-5, p. 14)* **8.** C *(SG 1-6, p. 17)* **9.** If you own a car, then you live in California. *(SG 1-6, p. 17)* **10.** integers from 0 to 100 *(SG 2-1, p. 25)* **11.** integers from 0 to 5 *(SG 2-1, p. 25)* **12.** $-\frac{1}{3}$ *(SG 2-2, p. 30)* **13.** A, C, D, and E *(SG 2-3, p. 35)* **14.** $\approx$4192 cm^3 *(SG 2-4, p. 38)* **15.** $d = 4.9t^2$ *(SG 2-5, p. 41)* **16.** 44.1 m *(SG 2-5, p. 41)* **17.** $\frac{1}{28x^2}$ *(SG 2-6, p. 45)* **18.** $\frac{3a^3}{2b^4}$ *(SG 2-6, p. 45)* **19.** −5 *(SG 2-6, p. 45)* **20.** $7xy^{1/2}$ *(SG 2-6, p. 45)* **21.** 2560 bacteria *(SG 2-7, p. 47)* **22.** (2, 3) *(SG 3-1, p. 52)* **23.** (7, −7) *(SG 3-2, p. 56)* **24.** no solution *(SG 3-3, p. 59)* **25.** (−1, 3) *(SG 3-4, p. 63)* **26.** $-\frac{2}{5}$ *(SG 3-3, p. 59)* **27.** $\begin{bmatrix} -3 & 3 \\ -1 & -2 \end{bmatrix}$ *(SG 3-5, p. 66)*

28. $\begin{bmatrix} 29 & 1 \\ -12 & 6 \end{bmatrix}$ *(SG 3-5, p. 66)* **29.** $\frac{3}{2}\begin{bmatrix} 2 & 4 & 3 \\ -2 & 4 & -6 \end{bmatrix}$ (columns A B C) *(SG 3-6, p. 70)*

$\begin{matrix} A' & B' & C' \end{matrix}$

30. $\begin{bmatrix} 3 & 6 & 4.5 \\ -3 & 6 & -9 \end{bmatrix}$ *(SG 3-6, p. 70)* **31.** a translation 4 units right and 5 units down *(SG 3-6, p. 70)*
32. $\begin{bmatrix} -7 & -7 \\ 12 & 4 \end{bmatrix}$ *(SG 3-7, p. 75)* **33.** [26 3 3] *(SG 3-7, p. 75)* **34.** $\begin{bmatrix} 1 & -3 \\ 4 & 1 \end{bmatrix}\begin{bmatrix} x \\ y \end{bmatrix} = \begin{bmatrix} 12 \\ 9 \end{bmatrix}$ *(SG 3-8, p. 78)*
35. (2, 11) *(SG 4-1, p. 84)* **36.** D *(SG 4-2, p. 88)*
37. E *(SG 4-2, p. 88)* **38.** B *(SG 4-2, p. 88)* **39.** C *(SG 4-2, p. 88)* **40.** A *(SG 4-2, p. 88)* **41.** 3, –3 *(SG 4-3, p. 91)* **42.** –1, 5 *(SG 4-4, p. 94)* **43.** –0.5, 2.5 *(SG 4-3, p. 91)* **44.** $(x^2 + 9)(x + 3)(x - 3)$ *(SG 4-4, p. 94)* **45.** –2.69, 0.19 *(SG 4-5, p. 97)* **46.** two distinct real roots *(SG 4-6, p. 99)* **47.** one real root *(SG 4-6, p. 99)* **48.** no real roots *(SG 4-6, p. 99)*
49. $7i, -7i$ *(SG 4-6, p. 99)* **50.** $1 - 5i$ *(SG 4-6, p. 99)*

Inventory 4

1. ≈102 *(SG 1-1, p. 1)* **2.** Spin the spinner 20 times, noting whether the arrow lands on the "raise two chicks" or "<2 chicks" on each spin. *(SG 1-2, p. 4)* **3.** convenience *(SG 1-3, p. 8)* **4.** cluster *(SG 1-3, p. 8)* **5.** Many shoppers will ignore the ballot and not bother to fill it out, so the response rate will be low. *(SG 1-4, p. 11)* **6.** The sum of the measures of the angles of a triangle is 180°. *(SG 1-5, p. 14)*
7. inductive reasoning *(SG 1-5, p. 14)* **8.** If a quadrilateral is a rectangle, then it has 4 right angles. *(SG 1-6, p. 17)* **9.** True. *(SG 1-7, p. 20)*
10. A *(SG 2-1, p. 25)* **11.** $-\frac{2}{3}$ *(SG 2-2, p. 30)*
12. 1.0; 1.5, 2.0, 4.0 *(SG 2-3, p. 35)* **13.** 41.6 cm^3 *(SG 2-4, p. 38)* **14.** $y = \frac{1}{8}x^2$ *(SG 2-5, p. 41)* **15.** $\frac{6}{b^3}$ *(SG 2-6, p. 45)* **16.** $\frac{2}{5}x^4y^2$ *(SG 2-6, p. 45)* **17.** 9 *(SG 2-6, p. 45)* **18.** $-2x^{1/3}y$ *(SG 2-6, p. 45)*
19. ≈63 years *(SG 2-7, p. 47)* **20.** $y < 2x + 2$; $y \geq -1$ *(SG 3-1, p. 52)* **21.** (3, 7) *(SG 3-2, p. 56)*
22. (–2, 3) *(SG 3-4, p. 63)* **23.** $y = \frac{1}{2}x + 2$ *(SG 3-3, p. 59)* **24.** $\begin{bmatrix} 2 & -1 & 6 \\ -6 & -2 & 7 \end{bmatrix}$ *(SG 3-5, p. 66)*
25. $\begin{bmatrix} 1.5 & -2.5 \\ 1 & 1.5 \end{bmatrix}$ *(SG 3-5, p. 66)* **26.** translation 2 units left and 3 units down *(SG 3-6, p. 70)*
27. dilation centered at the origin with scale factor $\frac{2}{3}$ *(SG 3-6, p. 70)* **28.** $\begin{bmatrix} 36 & 17 \\ 2 & 12 \end{bmatrix}$ *(SG 3-7, p. 75)*
29. $\begin{bmatrix} x \\ y \end{bmatrix} = \begin{bmatrix} 3 & -1 \\ -5 & 2 \end{bmatrix}\begin{bmatrix} 1 \\ 3 \end{bmatrix}$ *(SG 3-8, p. 78)* **30.** $\begin{bmatrix} 0 \\ 1 \end{bmatrix}$ *(SG3-8, p. 78)* **31.** $\left(\frac{3}{2}, -\frac{13}{4}\right)$ *(SG 4-1, p. 84)*
32. $y = -x^2 + 2$ *(SG 4-2, p. 88)* **33.** $y = (x + 3)^2$ *(SG 4-2, p. 88)* **34.** $y = (x - 1)^2 + 1$ *(SG 4-2, p. 88)*
35. $y = -(x - 1)^2$ *(SG 4-2, p. 88)* **36.** $3x(x + 2)(x - 4)$ *(SG 4-4, p. 94)* **37.** –4, 2 *(SG 4-3, p. 91)* **38.** $-\frac{1}{3}$ *(SG 4-4, p. 94)* **39.** 1.72; –0.38 *(SG 4-5, p. 97)*
40. two distinct real roots *(SG 4-6, p. 99)* **41.** no real roots *(SG 4-6, p. 99)* **42.** one real root *(SG 4-6, p. 99)* **43.** (3, 0) and (–1, 0) *(SG 4-7, p. 103)*
44. 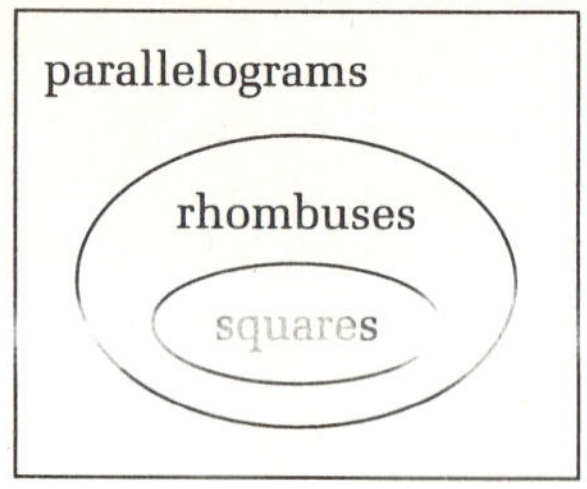

(SG 5-1, p. 108)

45. The slope of $\overleftrightarrow{AB}$ and $\overleftrightarrow{CD}$ is 2, while that of $\overleftrightarrow{BC}$ and $\overleftrightarrow{DA}$ is $-\frac{2}{5}$, so $\overline{AB}$ is parallel to $\overline{CD}$ and $\overline{BC}$ is parallel to $\overline{DA}$. *(SG 5-2, p. 111)* **46.** No. The four sides are not of equal length. $AB = \sqrt{5}$, while $BC = \sqrt{7.25}$. *(SG 5-2, p. 111)* **47.** (1.75, 0.5) *(SG 5-3, p. 114)* **48.** $A'(0, 0)$, $B'(1, -2)$, $C'(3.5, -1)$, $D'(2.5, 1)$ *(SG 5-4, p. 117)* **49.** $A(-a, 0)$, $B(0, b)$, $C(c, 0)$, $D(0, -b)$ *(SG 5-5, p. 121)* **50.** The diagonals are perpendicular. *(SG 5-6, p. 124)*

Inventory 5

1. 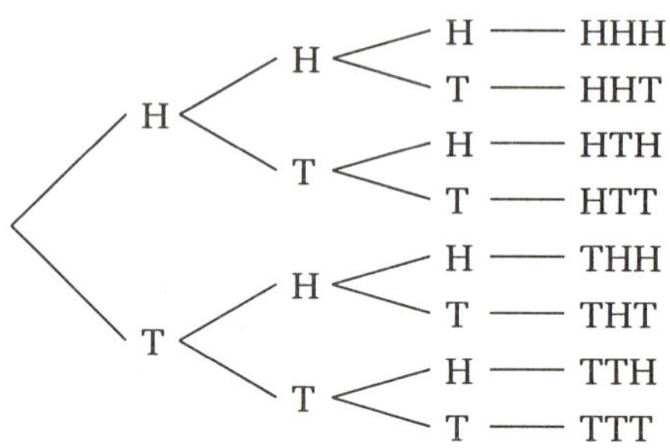

(SG 6-1, p. 130)
2. 18 outfits *(SG 6-2, p. 133)* **3.** 37% *(SG 6-3, p. 136)* **4.** $\frac{1}{6}$ *(SG 6-3, p. 136)* **5.** No. $P(\text{A and B}) = \frac{1}{5} \neq P(A)P(B) = \frac{1}{6}$ *(SG 6-4, p. 140)* **6.** $\frac{9}{38}$ *(SG 6-4, p. 140)* **7.** 120 *(SG 6-5, p. 143)* **8.** $2^8 = 256$ *(SG 6-6, p. 145)* **9.** ≈36% *(SG 6-8, p. 150)* **10.** $x^4 + 4x^3y + 6x^2y^2 + 4xy^3 + y^4$ *(SG 6-9, p. 153)*
11. Tomas called. *(SG 7-1, p. 158)* **12.** If $x^2 = 9$, then $x = 3$. *(SG 7-2, p. 162)* **13.** False. $(-3)^2$ is also equal to 9. *(SG 7-2, p. 162)* **14.** *or* rule *(SG 7-3, p. 165)* **15.** indirect argument *(SG 7-3, p. 165)*
16. True. Friday always follows Thursday and Thursday always comes before Friday. *(SG 7-4, p. 169)* **17.** False. Every triangle has at least one acute angle. An acute triangle has three acute angles. *(SG 7-4, p. 169)*

18. *Given:* $3x - 5 = 1$; *Prove:* $x = 2$

Statements: $3x - 5 = 1 \xrightarrow{❶} 3x = 6 \xrightarrow{❷} x = 2$

Justifications: ❶ Addition property of equality
❷ Division property of equality *(SG 7-6, p. 175)*

19. *Given:* $\angle 1$ and $\angle 2$ are supplementary, and $\angle 1$ is a right angle; *Prove:* $\angle 2$ is a right angle.

Statements	Justifications
1. 1 and 2 are suppl.	1. Given
2. $m\angle 1 + m\angle 2 = 180°$	2. Defn. of suppl. $\angle$s
3. $\angle 1$ is a right angle.	3. Given
4. $m\angle 1 = 90°$	4. Defn. of right $\angle$
5. $90° + m\angle 2 = 180°$	5. Substitution (steps 2 and 4)
6. $m\angle 2 = 90°$	6. Subtr. Prop. of =
7. $\angle 2$ is a right $\angle$	7. Defn. of right $\angle$

(SG 7-7, p. 177)

20. $\angle 1$ and $\angle 5$; $\angle 2$ and $\angle 6$; $\angle 3$ and $\angle 7$; $\angle 4$ and $\angle 8$ *(SG 7-8, p. 181)* **21.** 138°; 138°; 42°; 138° *(SG 7-8, p. 181)*

Inventory 6

1.

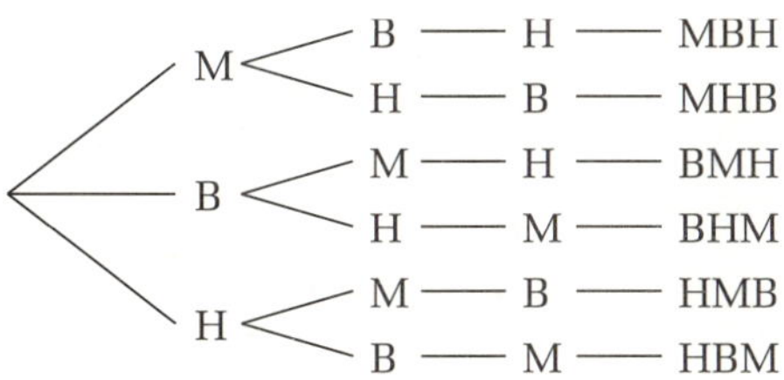

(SG 6-1, p. 130)

2. $6 \cdot 5 \cdot 4 \cdot 3 = 360$ *(SG 6-2, p. 133)* **3.** 45% *(SG 6-3, p. 136)* **4.** $\frac{25}{102} \approx 0.245$ *(SG 6-4, p. 140)* **5.** $\frac{1}{4}$ *(SG 6-4, p. 140)* **6.** 20,358,520 *(SG 6-5, p. 143)* **7.** ≈95% *(SG 6-8, p. 150)* **8.** 1 4 6 4 1 *(SG 6-6, p. 145)* **9.** $a^4 - 8a^3 + 24a^2 - 32a + 16$ *(SG 6-9, p. 153)* **10.**

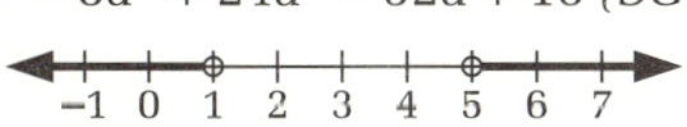

(SG 7-1, p. 158) **11.** If Raul gets a day off work, then he goes sailing. *(SG 7-2, p. 162)* **12.** I must take the bus. *(SG 7-3, p. 165)* **13.** A triangle is a right triangle if and only if it has a right angle. *(SG 7-4, p. 169)*

14. *Given:* $7x + 3 = -4$, *Prove:* $x = -1$

Statements: $7x + 3 = -4 \xrightarrow{❶} 7x = -7 \xrightarrow{❷} x = -1$

Justifications: ❶ Subtraction property of equality
❷ Division property of equality *(SG 7-6, p. 175)*

15. *Given:* Lines l and m are intersected by line t and $m\angle 1 = m\angle 3$, *Prove:* $m\angle 2 = m\angle 3$

Statements	Justifications
1. $m\angle 1 = m\angle 3$	1. Given
2. $\angle 1$ and $\angle 2$ are vertical angles	2. Defn. of vert. $\angle$s
3. $m\angle 1 = m\angle 2$	3. Vert. $\angle$s are = in measure
4. $m\angle 2 = m\angle 3$	4. Substitution (steps 1 and 3)

(SG 7-7, p. 177)

16. *Given:* Quadrilateral $ABCD$ is a parallelogram, *Prove:* $\angle A$ is supplementary to $\angle B$

Proof: Since $ABCD$ is a parallelogram, $\overleftrightarrow{AD} \parallel \overleftrightarrow{BC}$. These two parallel lines are intersected by transversal $\overleftrightarrow{AB}$. So, $\angle CBA$ and $\angle BAD$ are co-interior angles, and by the co-interior angle theorem, they are supplementary. *(SG 7-8, p. 181)*

17. No. *(SG 8-1, p. 188)* **18.** Yes. *(SG 8-1, p. 188)* **19.** $m\angle A = 30°$, $m\angle B = 50°$, $m\angle C = 100°$ *(SG 8-2, p. 192)* **20.** $m\angle B = m\angle E$ (given) and $m\angle BCA = m\angle DCE$, since they are vertical angles. So the triangles are similar by AA similarity postulate. *(SG 8-3, p. 198)* **21.** $AB = DE$ (AAS) or $BC = EC$ (ASA) *(SG 8-4, p. 203)* **22.** We are given that $AB = AC$ and $BD = CD$. By the reflexive property of equality we know that $AD = AD$. So $\triangle ADB \cong \triangle ADC$ by SSS. Then $m\angle BAD = m\angle CAD$, since corresponding parts of congruent triangles have equal measures. Thus, $\overrightarrow{AD}$ bisects $\angle A$. *(SG 8-5, p. 206)* **23.** $\triangle ABC \sim \triangle ACD \sim \triangle CBD$ *(SG 8-7, p. 215)* **24.** 13 *(SG 8-7, p. 215)* **25.** $x = 10\sqrt{3}$; $y = 20$ *(SG 8-8, p. 220)*
26. $a = b = 5$ *(SG 8-8, p. 220)*

Inventory 7

1.

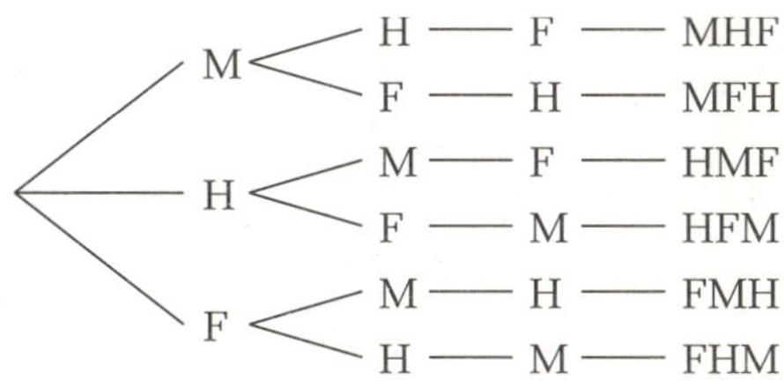

(SG 6-1, p. 130)

2. $5! = 120$ *(SG 6-2, p. 133)* **3.** $\frac{9}{14}$ *(SG 6-3, p. 136)*
4. 3:11 *(SG 6-3, p. 136)* **5.** $\frac{1}{20}$ *(SG 6-4, p. 140)*
6. 28.8% *(SG 6-4, p. 140)* **7.** 6188 committees *(SG 6-5, p. 143)* **8.** 1 6 15 20 15 6 1 *(SG 6-6, p. 145)*
9. ≈24% *(SG 6-8, p. 150)* **10.** $x^6 + 6x^5 + 15x^4 + 20x^3 + 15x^2 + 6x + 1$ *(SG 6-9, p. 153)* **11.** Rosa bought potato chips. *(SG 7-1, p. 158)* **12.** If you are in Cincinnati, then you are in Ohio. *(SG 7-2, p. 162)*
13. A spider is an arachnid. *(SG 7-3, p. 165)* **14.** If $2x + 4 = 8$, then $x = 2$. *(SG 7-3, p. 165)* **15.** It is not Consuela's birthday. *(SG 7-3, p. 165)* **16.** If a quadrilateral is a rectangle, then it has 4 right angles. If a quadrilateral has 4 right angles, then it is a rectangle. *(SG 7-4, p. 169)*

17. *Given:* $4x^2 = 36$ and $x < 0$. *Prove:* $x = -3$

Statements: $4x^2 = 36 \xrightarrow{❶} x^2 = 9 \xrightarrow{❷} \left.\begin{array}{r} x = \pm 3 \\ x < 0 \end{array}\right\} \xrightarrow{❸} x = -3$

Justifications: ❶ Division property of equality
❷ Definition of square root; Given ❸ *Or* rule
(SG 7-6, p. 175)

18. *Given:* Lines l and m intersected by line t; $m\angle 1 = m\angle 2$. *Prove:* $\angle 3$ and $\angle 2$ are supplementary.

Statements	Justifications
1. $\angle 1$ and $\angle 3$ together make a straight $\angle$.	1. Given
2. $m\angle 1 + m\angle 3 = 180°$	2. Straight $\angle$ postulate
3. $m\angle 1 = m\angle 2$	3. Given
4. $m\angle 2 + m\angle 3 = 180°$	4. Substitution (steps 2 and 3)
5. $\angle 2$ and $\angle 3$ are suppl. $\measuredangle$s	5. Defn. of suppl. $\measuredangle$s

(SG 7-7, p. 177)

19. $\overleftrightarrow{AB}$ and $\overleftrightarrow{CD}$; $\overleftrightarrow{AD}$ and $\overleftrightarrow{BC}$ *(SG 7-8, p. 181)* **20.** $(180 - x)°$; $x°$ *(SG 7-8, p. 181)* **21.** 360° *(SG 7-8, p. 181)* **22.** If two lines are ⊥ to the same transversal, then they are ‖. *(SG 8-1, p. 188)* **23.** When co-interior $\measuredangle$s are supplementary, the lines are parallel. *(SG 8-1, p. 188)* **24.** $a = 60°$, $b = 95°$ *(SG 8-2, p. 192)* **25.** $x = 105°$ *(SG 8-2, p. 192)* **26.** $a = 75°$ *(SG 8-2, p. 192)* **27.** $AC = 3\frac{1}{3}$; $BC = 3\frac{3}{5}$ *(SG 8-3, p. 198)* **28.** $\triangle ABC \cong \triangle DEC$; $AB = DE$, $m\angle B = m\angle E$, and $m\angle ACB = m\angle DCE$; therefore, the triangles are congruent by AAS. *(SG 8-4, p. 203)* **29.** $\triangle AMC \cong \triangle BMC$; $AM = BM$; $m\angle AMC = m\angle BMC$, and $MC = MC$; therefore, the triangles are congruent by SAS. *(SG 8-5, p. 206)* **30.** $\overline{BD}$ is the perpendicular bisector of $\overline{AC}$. *(SG 8-6, p. 210)* **31.** $3\sqrt{2}$ *(SG 8-7, p. 215)* **32.** 24 *(SG 8-8, p. 220)* **33.** $u = 4\sqrt{3}$; $v = 4$ *(SG 8-8, p. 220)* **34.** $\sin B = \frac{12}{13}$; $\cos B = \frac{5}{13}$; $\tan B = \frac{12}{5}$ *(SG 8-8, p. 220)* **35.** $170\pi = 2\pi r(3r + 2)$ *(SG 9-1, p. 227)* **36.** 4 *(SG 9-1, p. 227)* **37.** $x^4 - 5x^3 + 4x^2 + 12$ *(SG 9-1, p. 227)* **38.** 0 *(SG 9-1, p. 227)* **39.** $-64x^6y^3$ *(SG 9-2, p. 231)* **40.** $\frac{2b}{3a}$ *(SG 9-2, p. 231)* **41.** $3xy(2x - 1)(x + 1)$ *(SG 9-2, p. 231)* **42.** $2, -\frac{10}{9}$ *(SG 9-3, p. 234)* **43.** 3, −1 *(SG 9-3, p. 234)* **44.** 2, a triple zero *(SG 9-4, p. 238)* **45.** 0, −4, 1 *(SG 9-4, p. 238)* **46.** −3, a double zero, and $-\frac{1}{2}$ *(SG 9-4, p. 238)* **47.** Graph of $y = -3(x - 2)^3$ *(SG 9-4, p. 238)* **48.** Graph of $y = (2x + 1)(x + 3)^2$ *(SG 9-4, p. 238)* **49.** Graph of $y = x(x - 1)(x + 4)$ *(SG 9-4, p. 238)* **50.** $0, \frac{3}{2}, -4$ *(SG 9-5, p. 242)*

51.

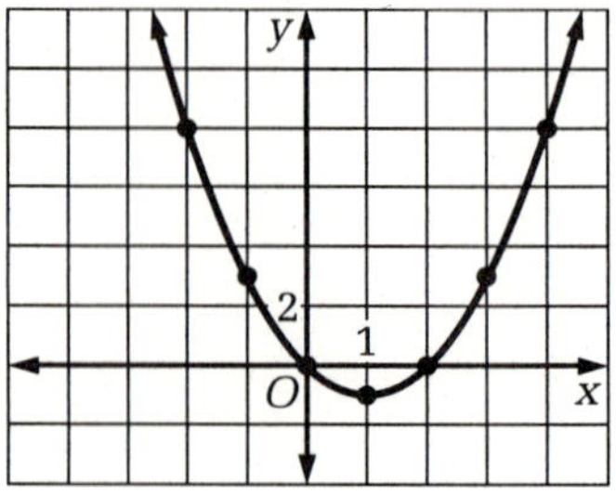

(SG 9-6, p. 244)